Justus Randolph

Computer Science Education Research at the Crossroads

Justus Randolph

Computer Science Education Research at the Crossroads

A Methodological Review of Computer Science Education Research

VDM Verlag Dr. Müller

Impressum/Imprint (nur für Deutschland/ only for Germany)
Bibliografische Information der Deutschen Nationalbibliothek: Die Deutsche Nationalbibliothek
verzeichnet diese Publikation in der Deutschen Nationalbibliografie; detaillierte bibliografische
Daten sind im Internet über http://dnb.d-nb.de abrufbar.
Alle in diesem Buch genannten Marken und Produktnamen unterliegen warenzeichen-, marken-
oder patentrechtlichem Schutz bzw. sind Warenzeichen oder eingetragene Warenzeichen der
jeweiligen Inhaber. Die Wiedergabe von Marken, Produktnamen, Gebrauchsnamen,
Handelsnamen, Warenbezeichnungen u.s.w. in diesem Werk berechtigt auch ohne besondere
Kennzeichnung nicht zu der Annahme, dass solche Namen im Sinne der Warenzeichen- und
Markenschutzgesetzgebung als frei zu betrachten wären und daher von jedermann benutzt
werden dürften.

Coverbild: www.purestockx.com

Verlag: VDM Verlag Dr. Müller Aktiengesellschaft & Co. KG
Dudweiler Landstr. 99, 66123 Saarbrücken, Deutschland
Telefon +49 681 9100-698, Telefax +49 681 9100-988, Email: info@vdm-verlag.de
Zugl.: Logan, Utah State University, 2007

Herstellung in Deutschland:
Schaltungsdienst Lange o.H.G., Berlin
Books on Demand GmbH, Norderstedt
Reha GmbH, Saarbrücken
Amazon Distribution GmbH, Leipzig
ISBN: 978-3-639-04425-6

Imprint (only for USA, GB)
Bibliographic information published by the Deutsche Nationalbibliothek: The Deutsche
Nationalbibliothek lists this publication in the Deutsche Nationalbibliografie; detailed
bibliographic data are available in the Internet at http://dnb.d-nb.de.
Any brand names and product names mentioned in this book are subject to trademark, brand or
patent protection and are trademarks or registered trademarks of their respective holders. The use
of brand names, product names, common names, trade names, product descriptions etc. even
without a particular marking in this works is in no way to be construed to mean that such names
may be regarded as unrestricted in respect of trademark and brand protection legislation and
could thus be used by anyone.

Cover image: www.purestockx.com

Publisher:
VDM Verlag Dr. Müller Aktiengesellschaft & Co. KG
Dudweiler Landstr. 99, 66123 Saarbrücken, Germany
Phone +49 681 9100-698, Fax +49 681 9100-988, Email: info@vdm-publishing.com
Copyright © 2008 VDM Verlag Dr. Müller Aktiengesellschaft & Co. KG and licensors
All rights reserved. Saarbrücken 2008

Printed in the U.S.A.
Printed in the U.K. by (see last page)
ISBN: 978-3-639-04425-6

Table of Contents

Chapter 1

———•———————————————————•———

Introduction

As technology comes to play an increasing role in the economic and social fabric of humanity, the need for computer science education will also increase. Computer science education can enable students to take part in a socio-technological future, help them understand the electronic world around them, and empower students to control, rather than be controlled by, technology. Furthermore, computer science education will help prepare students for higher education in the computing sciences and, consequently, help remedy the projected shortage of highly trained computing specialists required to keep economic infrastructures functional.

It is a given that research on computer science education can lead to the improvement of computer science education. However, computer science education research is acknowledged as being an emerging and isolated field. One way to improve an emerging field is with a review of the research methods used in that field so that those methods can be analyzed and improved upon.

In a methodological review, a content analysis approach is used to analyze the research practices reported in a body of academic articles. Methodological reviews differ from meta-analyses in that research practices, rather than research outcomes, are emphasized. They are known to be one way to improve the research methods of a field because they provide a solid basis on which to make recommendations for improvements in practice. Methodological reviews have been successfully used to inform policy and practice in fields such as educational technology and behavioral science statistics.

Although there have been methodological reviews of computer science education research, they have either examined nonrepresentative samples of research articles or have been of poor quality. Because of the benefits that can be reaped from methodological reviews and because the previous methodological reviews of computer science education research are lacking, I conducted a rigorous methodological review, from a behavioral science perspective, on a representative sample of all the research articles published in major computer science education research journals and conference proceedings from 2000-2005.

———————————————————————————
1

I expect that this text will make a contribution to the field by supplying a solid ground on which to make recommendations for improvement and to promote informed dialogue about computer science education research. If my recommendations are heeded, I expect that computer science education research will improve, which will improve computer science education, which will, in turn, help the technologically oriented social and economic needs of the future be met.

The Importance of Computer Science Education

The study of the discipline of computing, defined as "the systematic study of algorithmic proceses that describe and transform information: their theory, analysis, design, efficiency, implementation, and application" (Denning et al., 1989, p. 12) is considered to be a key factor in preparing K-12 students, and people in general, for a technologically oriented future (see Tucker et al., 2003, p. 4). (In this book I use the term *computer science education*, rather than the more general term computing education, since *computer science education* is the term adopted by the Association for Computing Machinery's Special Interest Group on Computer Science Education.) The National Research Council Committee on Information Technology Literacy (NRC; 1999) provides strong rationales for teaching students about technology and computer science. The NRC argues that people will increasingly need to understand technology to carry out personally meaningful and necessary tasks, such as

- Using e-mail to stay in touch with family and friends
- Pursuing hobbies
- Helping children with homework and projects
- Finding medical information or information about political candidates over the World Wide Web (n.p.)

The NRC also argues that an informed citizenry must also be a citizenry that has a high degree of technological fluency because many contemporary public policy debates are associated with information technology. For example, the NRC wrote,

> A person with a basic understanding of database technology can better appreciate the risks to privacy entailed in data-mining based on his or her credit card transactions. A jury that understands the basics of computer animation and image manipulation may have a better understanding of what counts as "photographic truth" in the reconstruction of a crime or accident. . . . A person who understands the structure and operation of the World Wide Web is in a better position to evaluate and appreciate the policy issues related to the First

Amendment, free expression, and the availability of pornography on the Internet. (n.p.)

In terms of U.S. labor needs, the U.S. Department of Commerce's Office of Technology Policy found that there was "substantial evidence that the United States is having trouble keeping up with the demand for new information technology workers" (as cited in Babbit, 2001, p. 21). *Computer support specialist* and *systems administrator* are expected to be two of the fastest growing U.S. occupations during the decade from 2002 to 2012 (U.S. Department of Labor-Bureau of Labor Statistics, n.d.a). Also, employment for computer systems analysts, database administrators, and computer scientists "is expected to increase much faster than average as organizations continue to adopt increasingly sophisticated technologies" (U.S. Department of Labor-Bureau of Labor Statistics, n.d.b).

Computer Science Education Research Can Improve Computer Science Education

Researchers, such as Gall, Borg, and Gall (1996), have shown that education research contributes to the practice of education. Gall and colleagues argue that educational research contributes four types of knowledge to the field of education—description, prediction, improvement, and explanation—and that education research enables practitioners to use "research knowledge about what *is* to inform dialogue about what *ought* to be" (p. 13). They further claim that basic educational research has been shown to influence practice even when influencing practice was not its intention.

If Gall and colleagues (1996) are correct, in as much as computer science education is a subset of education research proper, then computer science education research also has the potential to make contributions to computer science education. However, as I argue in the section below, currently the realization of that potential is uncertain; there needs to be more research knowledge about what is to inform what ought to be.

Computer Science Education Research Is an Isolated, but Emerging Field

The seminal book on computer science education research (Fincher & Petre, 2004) begins with the following statement: "Computer science education research is an emergent area and is still giving rise to a literature" (p. 1). Top computer science education researchers like Mark Guzdial and Vicki Almstrum argue that the interdisciplinary gap between computer science education and educational research proper, including methods developed in the broader field of behavioral research, must be overcome before computer science education research can be considered to be a field which has emerged (Almstrum, Hazzan, Guzdial, & Petre, 2005). (In this book, I use the term *behavioral research* as a synonym for what Guzdial, in Almstrum et al. [2005, p. 192], calls "education, cognitive science, and learning sciences research.") Addressing this lack of connection with behavioral research, Guzdial, in Almstrum and colleagues (2005), wrote:

> The real challenge in computer education is to avoid the temptation to re-invent the wheel. Computers are a revolutionary human invention, so we might think that teaching and learning about computers requires a new kind of education. That's completely false: The basic mechanism of human learning hasn't changed in the last 50 years.
> Too much of the research in computing education ignores the hundreds of years of education, cognitive science, and learning sciences research that have gone before us. . . . If we want our research to have any value to the researchers that come after us, if we want to grow a longstanding field that contributes to the improvement of computing education, then we have to "stand on the shoulders of giants," as Newton put it, and stop erecting ant hills that provide too little thought. (pp. 191-192)

The findings from three previous methodological reviews—(a) a critical review of Kindergarten through 12th-grade (K-12) computer science education program evaluations, (b) a methodological review of selected articles published in the SIGCSE Technical Symposium Proceedings, and (c) a methodological review of the full-papers published in the Proceedings of the Koli Calling Conference on Computer Science Education triangulate to support the idea that computer science education research and evaluation is indeed an emerging and isolated field. (In this book, by *program*, I mean a *project*, not *software*.) The findings from those three previous reviews (i.e., Randolph, 2005; Randolph, Bednarik, & Myller, 2005; Valentine, 2004) are summarized below.

A Methodological Review of K-12 Computer Science Education Program Evaluations

Earlier I conducted a methodological review and meta-analysis of the program evaluation reports in computer science education, which is reported in Randolph (2005). (Throughout this book, because of the difficulties of making an external decision about what is *research* and what is *evaluation*, I operationalize an *evaluation report* as a document that the author called an *evaluation, evaluation report,* or *a program evaluation report*.) To identify the strengths and weaknesses in K-12 computer science education program evaluation practice, I attempted to answer the following questions:

1. What are the methodological characteristics of computer science education program evaluations?
2. What are the demographic characteristics of computer science education evaluation reports?
3. What are the evaluation characteristics of computer science education program evaluations?
4. What is the average effect of a particular type of program on computer science achievement?

Electronic searches of major academic databases, the Internet, and the ACM digital library; a subsequent reference-branching hand search; and a query to over 4,000 computer science education researchers and program evaluators were the search techniques used to collect a comprehensive sample of K-12 computer science education program evaluations. After selecting the evaluation reports that met seven stringent criteria for inclusion, the sample of program evaluations were then coded in four areas: demographic characteristics, intervention characteristics, evaluation characteristics, and findings. In all, 14 main variables were coded for: region of origin, source, decade of publication, grade level of target participants, target population, area of computing curriculum, program activities, outcomes measured, moderating factors examined, measures, type of measures, type of inquiry, experimental design, and study quality. Additionally, Cohen's d was calculated for impact on computer science achievement for each study that reported enough information to do so. A second rater coded a portion of the reports on the key variables to estimate levels of interrater reliability.

Frequencies and percents were calculated for each of the variables above. A random effects, variance and within-study sample size/study-quality weighting approach was used to combine effect sizes. Interactions were examined for type of program.

In all, 29 evaluation reports were included. Eight of those reports had data that could be converted to effect sizes and were included in the meta-analytic portion of the article, where the effect sizes were synthesized. The major findings are summarized below:

1. Most of the programs that were evaluated offered direct computer science instruction to general education, high school students in North America.
2. In order of decreasing frequency, evaluators examined stakeholder attitudes, program enrollment, academic achievement in core courses, and achievement in computer science.
3. The most frequently used measures were, in decreasing order of frequency, questionnaires, existing sources of data, standardized tests, and teacher-made or researcher-made tests.

Only one measure of computer science achievement, which is no longer available, had reliability or validity estimates. The pretest-posttest design with a control group and the one-group posttest-only design were the most frequently used research designs.

4. No interaction between type of program and computer science achievement improvement was found.

In terms of the link between program evaluation and computer science education, the fact that there were so few program evaluations being done, that so few of them (i.e., only eight) went beyond simple program description and student attitudes, that only one used an instrument with information about measurement reliability and validity, and that one-group posttest-only designs were so frequently used indicate that the past K-12 computer science education program evaluations have had many deficiencies. As the next review indicates, the deficiencies are not solely found in K-12 computer science education program evaluations; there are also several deficiencies in K-12 computer science education research as well.

A Methodological Review of Selected Articles in
SIGCSE Technical Symposium Proceedings

Valentine (2004) critically analyzed over 20 years of computer science education conference proceedings that dealt with first-year university computer science instruction. In that review, Valentine categorized 444 articles into six categories. The major finding from his review was that only 21% of papers in the last 20 years of proceedings were categorized as *experimental*, which was operationalized as the author of the paper making "any attempt at assessing the 'treatment' with some scientific analysis" (p. 256). Some of Valentine's other findings are listed below:

1. The proportion of experimental articles had been increasing since the mid-90s.
2. The proportion of what he calls *Marco Polo—I went there and I saw this—* types of papers had been declining linearly since 1984.
3. The overall number of papers being presented in the SIGCSE forum had been steadily increasing since 1984 (as cited in Randolph, Bednarik, & Myller, 2005, p. 104).

Valentine concluded that the challenge is to increase the number of experimental investigations and decrease the number of "I went there and saw that," self-promotion, or descriptions-of-tools types of articles. The reliability of Valentine's findings, however, is questionable; Valentine was the single coder and reported no estimates of interrater agreement.

A Methodological Review of the Papers Published
in Koli Calling Conference Proceedings

Randolph, Bednarik, and Myller (2005) conducted a critical, methodological review of all of the full-papers in the proceedings of the *Koli Calling: Finnish/Baltic Sea Conference on Computer Science Education* (hereafter *Koli Proceedings)* from 2001 to 2004. Each paper was analyzed in terms of (a) methodological characteristics, (b) section proportions (i.e., the proportion of literature review, methods, and program description sections), (c) report structure, and (d) region of origin. Based on an analysis of all of the full-papers published in four years of Koli Proceedings, their findings were that

1. The most frequently published type of paper in the Koli Proceedings was the program (project) description.
2. Of the empirical articles reporting research that involved human participants, exploratory descriptive (e.g., survey research) and quasi-experimental methods were the most common.
3. The structure of the empirical papers that reported research involving human participants deviated sharply from structures that are expected in behavioral science papers. For example, only 50% of papers that reported research on human participants had literature reviews; only 17% had explicitly stated research questions.
4. Most of the text in empirical papers was devoted to describing the evaluation of the program; very little was devoted to literature reviews.
5. The Koli Calling proceedings represented mainly the work of Nordic/Baltic, especially Finnish, computer science education researchers.
6. An additional finding was that no article reported information on the reliability or validity of the measures that were used.

Both the Valentine (2004) and Randolph, Bednarik, and Myller (2005) reviews converged on the finding that few computer science education research articles went beyond describing program activities. In the rare cases when impact analysis was done, it was usually done using anecdotal evidence or with weak research designs.

Synthesis of Findings across Methodological Reviews

When synthesizing the results of these methodological reviews, between methodological reviews, several preliminary themes from the papers covered in the methodological reviews emerged. They are listed below:

1. There is a paucity of impact evaluation/research.
2. There is a proliferation of purely program descriptions.
3. There is an urgent need for reliable and valid measures of computer science achievement.
4. Research/evaluation reports concentrate mainly on stakeholder attitudes towards a computer science education program.
5. When experiments or quasi-experiments are conducted, the research designs are weak.
6. There is a huge gap between how research on human participants is conducted and reported by computer-science-grounded practitioners and by behavioral-science-grounded practitioners. (Even the term *evaluation* is used differently by practitioners in these two groups. See Randolph & Hartikainen, 2004.)
7. Literature reviews in computer science education research and evaluation reports are missing or inadequate.

Table 1 shows which review provided evidence for each of the themes listed above. In essence, the findings of the three reviews described above do in fact converge on Fincher and Petre's (2004) hypothesis that computer science education research is an emerging, but isolated, field.

Methodological Reviews Can Improve Research Practice

In many types of literature reviews the emphasis is on the analysis and integration of research outcomes and on how study characteristics covary with outcomes. In fact, the ERIC Processing Manual defines "a literature review" as an "information analysis and synthesis, focusing on outcomes . . ." (as cited in Cooper & Hedges, 1994, p. 4). In methodological reviews, however, the emphasis is not on research outcomes, but on the description and analysis of research practices (see Cooper, 1988). Keselman et al. (1998) wrote,

Table 1. *Evidence Table for Themes of the Literature Review*

Theme	Randolph, 2005	Valentine, 2004	Randolph, Bednarik, & Myller, 2005
Paucity of impact research	x	X	X
Mostly program descriptions	x	X	X
Need for measures	x		X
Stakeholder attitudes	x		X
Weak designs	x		X
Research gap	x		X
Lack of lit. reviews			X

> Reviews typically focus on summarizing the results of research in particular areas of inquiry (e.g., academic achievement of English as a second language) as a means of highlighting important findings and identifying gaps in the literature. Less common, but equally important, are reviews that focus on the research process, that is, the methods by which a research topic is addressed, including research design and statistical analyses issues. (pp. 350-351)

As an example, in a methodological review of educational researchers' ANOVA, MANOVA, and ANCOVA analyses, Keselman and colleagues (1998) used a content analysis approach to synthesize the statistical practices in research articles published in major education research journals. They then compared the statistical practices of educational researchers with the statistical practices recommended by statisticians and made recommendations for improvement.

Of the variety of reasons for conducting a methodological review, two of the most obvious reasons are to help improve methodological practice and inform editorial policy. According to Keselman and colleagues (1998),

> Methodological reviews have a long tradition (e.g., Edgington, 1964; Elmore & Woehlke, 1988, 1998; Goodwin & Goodwin, 1985a, 1985b; West, Carmody, & Stallings, 1983). One purpose of these reviews had been the identification of trends in . . . practice. The

documentation of such trends has a twofold purpose: (a) it can form the basis for recommending improvement in research practice, and (b) it can be used as a guide for the types of . . . procedures that should be taught in methodological courses so that students have adequate skills to interpret the published literature of a discipline and to carry out their own projects. (pp. 350-351)

One current example of how methodological reviews can bring about improved practice and inform policy is shown in Leland Wilkinson and the APA Task Force on Statistical Inference's influential 1999 report—*Statistical Methods in Psychology Journals: Guidelines and Explanations* (hereafter *Wilkinson et al*). In that report, several of the most prominent figures in behavioral science research (e.g., Robert Rosenthal, Jacob Cohen, Donald Rubin, Bruce Thompson, Lee Cronbach, and others) came together, in response to the use and abuse of inferential statistics reported in Cohen (1994), to codify best practices in inferential statistics and in statistical methods in general. In that report, they drew heavily on methodological reviews of the statistical practices of behavioral science researchers, such as Keselman and colleagues (1998), Kirk (1996), and Thompson and Snyder (1998). Keselman and colleagues were interested in the ANOVA, ANCOVA, and MANOVA practices used by educational researchers. Kirk and Thompson and Snyder were interested in the statistical inference and reliability analyses done by education researchers. In addition to the fields of psychological statistics, methodological reviews have also been published in other fields, from program evaluation (Lawrenz, Keiser, & Lavoir, 2003; Randolph, 2005) to special education (Test, Fowler, Brewer, & Wood, 2005) to medical science (Clark, Anderson, & Chalmers, 2002; Huwiler-Müntener, Jüni, Junker, & Egger, 2002; Lee, Schotland, Bacchetti, & Bero, 2002).

In general terms, The Social Science Research Council (SSRC) and the National Academy of Education's (NAE) Joint Committee on Education Research noted a lack of and need for "data and analysis of the education research enterprise" (Ranis & Walters, 2004, p. 798). In fact the research priorities concerning the lack of data and analysis in education research included "determination of where education research is conducted and by whom" and "identification of the range of problems addressed and the methods used to address them" (p. 799). Methodological reviews can help meet the need for data about and analysis of the education research enterprise, especially regarding the research priorities identified above.

There are two conditions that suggest the value for a methodological review to improve practice and inform policy. The first is when there is consensus among experts for "best practice" but actual practice is expected to fall far short of best practice. The methodological review can identify these shortcomings and suggest policies for research funding and publication. For example, in the Keselman and colleagues (1998) review, they found that there was a difference between how statisticians use ANOVA and how social science researchers use ANOVA. Thus, the rationale for the Keselman and colleagues review was that the recommendations given by the statisticians could benefit the research practices of the social science researchers. The second condition is when there are islands of practice that can benefit from exposure to each other—for example, when there are groups that practice research in different ways or at different levels.

In terms of the conditions for a methodological review to improve practice and inform policy, both conditions are met for the field of computer science education. First, there are islands of practice. As Guzdzial points out in the statement of the Association for Computing Machinery's Special Interest Group on Computer Science Education's (hereafter *ACM SIGCSE*) panel on 'Challenges to Computer Science Education Research,' there are two distinct islands of practice: computer science education research and "education, cognitive science, and learning sciences research" (Almstrum et al., 2005, p. 192). Second, there is a call for interdisciplinary exchange between islands of practice; actual practice in computer science education research differs from accepted practice in "education, cognitive science and learning sciences research." The ACM SIGCSE panel on 'Challenges to Computer Science Education Research' stated that one of the keys to improving computer science education research is for computer science educators to look to "education, cognitive science, and learning sciences research." This sentiment was also stated by the computer science education panel on Import and Export to/from Computing Science Education (Almstrum, Ginat, Hazzan, & Morely, 2002). They wrote:

> Computing science education is a young discipline still in search of its research framework. A practical approach to formulating such a framework is to adapt useful approaches found in the research from other disciplines, both educational and related areas. At the same time, a young discipline may also offer innovative approaches to the older discipline. (p. 193)

Methodological Reviews in the Field of Educational Technology

Psychology is not the only field in which methodological reviews have been conducted. The field of educational technology, which makes use of the software engineering and management information systems components of computer science, has a long history of methodological reviews, dating as far back as the mid-1970s. To make sense of all of those reviews and to be able to compare the results of this book across fields, I conducted a review of those methodological reviews. Specifically, I attempted to answer the following research questions:

1. What metacategories can be used to subsume the categories used in the previous educational technology methodological reviews?
2. What proportions of articles in the previous educational technology methodological reviews fall into each of these categories?
3. How do those proportions of articles differ by year and type of forum?
4. How do these proportions compare with the proportions in education research proper?

In the sections that follow I (a) present the results of a methodological review of education proper articles (to be able to answer Question 4), (b) present the methods for conducting this review of methodological reviews of education technology articles, and (c) finally present the results of the review of methodological reviews of educational technology articles.

The Proportions of Article Types in Education Research Proper

Before describing the methods that were used in this review of reviews, to have a point of reference on which this review's results can be compared and contrasted, I report on a high-quality methodological review in the field of education research proper. In that review, Gorard and Taylor (2004) reviewed 42 articles from the six issues published in 2001 in the *British Educational Research Journal* (BERJ), 28 articles from the four issues published in 2002 in the *British Journal of Educational Psychology* (BJEP), and 24 articles from the four issues published in 2002 in *Educational Management and Administration*. Gorard and Taylor found the following results:

> Overall, across three very different [education] journals in 2002, 17 per cent of articles were clearly or largely non-empirical (although this description includes literature reviews, presumably based on empirical evidence), 4 percent were empirical pieces using a combination of 'qualitative' and 'quantitative' methods (therefore a rather rare phenomenon), 34 percent used qualitative methods alone, and 47 percent used quantitative methods alone. (p. 141)

Because the cumulative percent above is 102, I rounded some figures down and assumed then that, out of 94 articles, 15, 4, 32, and 43 articles were nonempirical, mixed, qualitative, and quantitative, respectively.

Although Gorard and Taylor's (2004) sample of articles that were reviewed was small, Gorard and Taylor provided convincing evidence, from a variety of sources, that validated the proportions of nonempirical, quantitative, qualitative, and mixed-methods articles found in their review. Those sources included

- interviews with key stakeholders from across the education field, includingresearchers, practitioner representatives, policy makers and policy implementers;
- a large-scale survey of the current methodological expertise and future training needs of UK education researchers; [and a]
- detailed analysis and breakdown of 2001 RAE [Research Assessment Exercise, 2001]. (p. 114)

Method for Conducting a Review of Methodological Reviews

In this section I explain the methods that I used for conducting this review of methodological reviews in educational technology. It includes a description of the criteria for inclusion and exclusion, the search strategy, coding categories, and data analysis procedures.

Criteria for Inclusion and Exclusion

Articles were included in this review if they met six criteria, which are listed below:

1. It was a quantitative review (e.g., a content analysis) of research practices, not a literature review in general or a meta-analysis, which focuses on research outcomes.
2. The review dealt with the field of educational technology or distance education.
3. The review was written in English.
4. The number of articles that were reviewed was specified.
5. The candidate review's categories were able to be subsumed under metacategories.
6. The review's articles did not overlap with another review's articles. (When reviews overlapped, only the most comprehensive review was taken.)

Search Strategy

The first step of the search strategy was to conduct an electronic search of the academic databases Academic Search Elite, Psych Info, and ERIC, and of the Internet, via Google. The electronic search was conducted in July 2006 using the terms *educational technology, methodological review; computer-assisted instruction, methodological review; educational technology, review;* and *computer-assisted instruction, review.* The title of each entry was read to determine if it might lead to a review that would meet the criteria for inclusion. (In cases where the review returned more than 500 entries, only the

---

first 500 were read.) If the title looked promising, the resulting webpage, abstract, or entire article was read to see if the article met the criteria for inclusion.

The second step of the search strategy was to do pearl building. The references section of the articles identified from the electronic search and the articles that were known to me beforehand were searched. This pearl-building process was repeated until a point of saturation was reached.

The third step of the search strategy was to compile a list of articles that met the criteria for inclusion and to send that list out to experts in the field of educational technology to see if there were any methodological reviews that should have been included on the preliminary list but had not. A query was sent to the members of the ITFORUM listserv on July 20, 2006. Eight ITFORUM members responded to the query and suggested more articles that might meet the criteria for inclusion.

Coding Categories

Each of the methodological reviews that met all six criteria was coded on seven attributes:

1. The forum from which the methodological review came;
2. The author(s) of the methodological review;
3. The year of the methodological review;
4. The forums, issues, and time periods from which the reviewed articles came;
5. The categories that each methodological review used;
6. The number of articles that were put into each of the methodological review's categories; an
7. The research question that the review attempted to answer.

Data Analysis

In the reviews which met all six criteria for inclusion, the number of articles which fit into each metacategory was recorded. Those results were summed to arrive at an overall picture of how many articles, across methodological reviews, fell into each of the metacategories. Those results were disaggregated by forum and by year. Also, the results of this methodological review of articles from educational technology forums were compared with the results of Gorard and Taylor's (2004) methodological review of articles from education research journals proper. Chi-square analyses were used to determine the likelihood of getting differences in the observed multinomial proportions as large as those expected by chance. In addition to the quantitative synthesis, I also recorded the research question that each methodological review attempted to answer.

Results of the Review of Reviews

The literature search resulted in 13 methodological reviews that met at least the first three criteria for inclusion (Alexander & Hedberg, 1994; Caffarella, 1999; Clark & Snow, 1975; Dick & Dick, 1989; Driscoll & Dick, 1999; Higgins, Sullivan, Harper-Marinick, & Lopez, 1989; Klein, 1997; Phipps & Merisotis, 1999; Randolph, in press; Randolph, Bednarik, Silander, et al., 2005; Reeves, 1995; Ross & Morrison, 2004; Williamson, Nodder, & Baker, 2001). Four of the reviews mentioned above did not meet all six criteria for inclusion and, therefore, were not included in the current review. Phipps and Merisotis's review, a large scale critical review of the research on distance learning, was excluded because it did not meet Criterion 4: it did not specify how many articles were reviewed. Ross and Morrison's review and Alexander and Hedberg's review were excluded because they did not meet criterion five: Ross and Morrison categorized by experimental design and setting, Alexander and Hedberg categorized by evaluation design. Also, Caffarella, who did a review of educational technology dissertations, was excluded because the categories used could not be codified with the metacategories in the current review. Driscoll and Dick was excluded because their sample overlapped with Klein's review, which had a more comprehensive sample. Reeves' sample of articles from *Educational Technology Research & Development* was not included because it also overlapped with Klein's review; however, Reeve's sample of *Journal of Computer-Based Instruction* articles was included. Thus, nine methodological reviews, covering 905 articles from the last 30 years of educational technology, were included in this review of educational technology methodological reviews. The questions that each of those methodological reviews attempted to answer are summarized in Table 2. At a glance, the question being asked in the major methodological reviews of the educational technology literature was "What are the types and methodological properties of research reported in educational technology articles?"

Table 3 presents those reviews, the forum, the years sampled, and the number of articles reviewed. As shown in Table 3 the forums that were covered in the previous reviews were *AV Communication Review* (AVCR), *Educational Communication and Technology Journal* (ECTJ), *Journal of Instructional Development* (JID), *Journal of Computer-Based Instruction* (JCBI), *Educational Technology Research & Development* (ETR&D), *American Journal of Distance Education* (AJDE), *Distance Education* (DE), *Journal of Distance Education* (JDE), *Proceedings of the International Conference on Advanced Learning Technologies* (ICALT). Also, of the 46 papers reviewed in Williams et al. (2001), Williams et al. wrote the following:

Table 2. *Research Questions in Educational Technology Methodological Reviews*

Study	Overview of research questions
Alexander & Hedberg, 1994	What, and in what proportions, evaluation models are used in evaluations of educational technology?
Caffarella, 1999	How have the themes and research methods of educational technology dissertations changed over the past 22 years?
Clark & Snow, 1975	What research designs are being reported in educational technology journals? In what proportions?
Dick & Dick, 1989	How do the demographics, first authors, and substance of articles in two certain educational technology journals differ?
Driscoll &Dick, 1999	What types of inquiry are being reported in educational technology journals? In what proportions?
Klein, 1997	What types of articles and what topics are being published in a certain educational technology journal? In what proportions?
Higgins et al., 1999	What do members of a certain educational technology journal want to read?
Phipps & Merisotis, 1999	What are the methodological characteristics of studies published in major educational technology forums?
Randolph, in press	Are the same methodological deficiencies reported in Phipps & Merisotis (1999) still present in current research?
Randolph et al., 2005	What are the methodological properties of articles in the proceedings of ICALT 2004?
Ross & Morrison, 2004	What are proportions of experimental designs being used in educational technology research?
Reeves, 1995	What types of methodological orientations do published educational technology articles take? In what proportions?
Williamson et al., 2001	What types of research methods and pedagogical strategies are being reported in educational technology forums?

Table 3. *Characteristics of Educational Technology Reviews Included in the Quantitative Synthesis*

Review	Forum	Years covered	Number of articles reviewed
Clark & Snow, 1975	AVCR	1970-1975	111
Dick & Dick, 1989	ECTJ	1982-1986	106
	JID	1982-1986	88
Higgins et al., 1989	ECTJ	1986-1988	40
	JID	1986-1988	50
Reeves, 1995	JCBI	1989-1994	123
Klein, 1997	TR&D	1989-1997	100
Williamson et al., 2001	Mixed	1996-2001	46
Randolph, in press	AJDE	2002	12
	DE	2002	14
	JDE	2002-2003	40
Randolph, 2005	ICALT	2004	175[a]
Total			905

Note. AVCR = *Audio Visual Communication Review,* ECTJ = *Educational Communication and Technology Journal,* JID = *Journal of Instructional Development,* JCBI = *Journal of Computer-Based Instruction,* ETRD = *Educational Technology Research & Development,* AJDE = *American Journal of Distance Education,* DE = *Distance Education,* JDE = *Journal of Distance Education,* ICALT = *International Conference on Advanced Learning Technologies.*

[a] 175 investigations reported in 123 articles

37 originate[d] from refereed journals or conference proceedings and the remainder from academic websites or Government departments. . . . In particular we drew material from the conferences of the Australasian Society for Computers in Learning in Tertiary Education (ASCILITE) and from the National Advisory Committee for Computing Qualifications (NACCQ). (p. 568)

Table 4 shows the categories that were used in previous methodological reviews. It shows how I grouped these categories together to arrive at the four metacategories: *qualititative*, *quantitative*, *mixed-methods,* and *other*. The *other* category included articles that did not deal with human participants, such as literature reviews, descriptions of tools, or theoretical papers.

Figure 1 shows the number and percentage of 905 articles that were distributed into each metacategory. The *other* category is the largest category, *experimental* is the second largest category, and those categories are followed by the *qualitative* and *mixed methods* categories.

Figure 2 shows the proportions of articles that fell into each of the different categories in each forum. It indicates that there as considerable variability between forums in terms of the proportions of types of articles that were published. It should be noted that these data usually only represent a limited time span over the life of the forum.

Figure 3 shows that the proportions of types of articles varied over each time period. (Note that the *other* category was not included here so the remaining categories could be more easily compared.) This figure shows that there were high proportions of qualitative articles from the early 80s to early 90s, but the proportions dropped off in the late 90s and early 00s. It is important to note when interpreting Figure 3 that forums were not constant across time periods; some forums were sampled more heavily in different time periods than others. Table 3 showed how many articles were sampled from each forum each time period. The median year in a yearly range determined what time period a review would be categorized into.

Table 5 shows the difference between the numbers of articles dealing with human participants in the current review of educational technology reviews and Gorard and Taylor's (2004) methodological review of British educational research. In short, education proper articles had, on average, 30% more articles that reported research on human participants than in educational technology articles. The difference was statistically significant, $\chi^2(1, N = 999) = 30.21, p < .000$. Table 6 shows, however, that the proportions of quantitative, qualitative, and mixed-methods articles were nearly the same in educational technology and general education-research forums. The differences were *not* statistically significant, $\chi^2(2, N = 573) = 1.41, p = .495$.

Table 4. *The Composition of Educational Technology Metacategories*

Qualitative	Quantitative	Mixed methods	Other
Qualitative; critical theory; explanatory descriptive; case studies	Quantitative; experimental/quasi-experimental; quasi-experimental; exploratory descriptive, correlational; causal-comparative; classification; descriptions; experimental research; experimental study; survey research, empirical research; evaluation; correlational; empirical, experimental, or evaluation; quantitative descriptive	Mixed methods; triangulated; mixed	Literature reviews; other; description with no data; theory, position paper, and so forth.; theory; methodology; professional

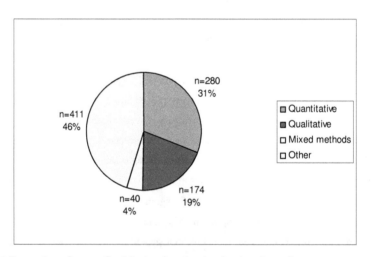

Figure 1. Proportion of types of articles in educational technology journals.

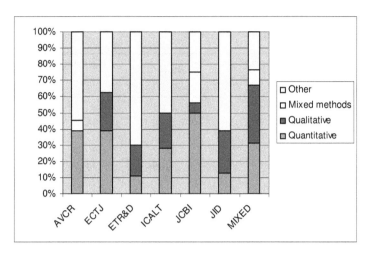

Figure 2. Proportion of types of educational technology articles by forum.

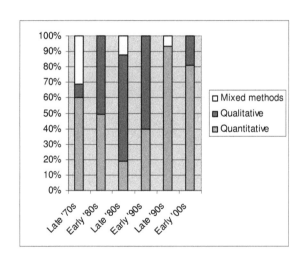

Figure 3. Proportions of types of educational technology articles by time period.

20

Table 5. *Comparison of the Proportion of Human Participants Articles in Educational Technology and Education Proper*

Field	Human participants		Total	Percentage yes	Adjusted residual
	Yes	No			
Ed. tech	494	411	905	54.6	-5.5
Ed. proper	79	15	94	84.0	5.5
Total	573	426	999		

Note. Ed. tech. = educational technology, Ed. proper = education proper $\chi^2(1, N = 999) = 30.21$, $p < .000$.

Table 6. *Comparison of Type of Methods Used in Educational Technology and Education Proper*

Type of article	Field		Total
	Ed. tech	Ed. proper	
Quantitative	280 (56.7%)	43 (54.4%)	323 (56.4%)
Qualitative	174 (35.2%)	32 (40.5%)	206 (36.0%)
Mixed methods	40 (8.1%)	4 (5.1%)	44 (7.7%)
Total	494 (100%)	79 (100%)	573 (100%)

Note. Percentages are within Review; Ed. tech. = educational technology. Ed. proper = education proper. $\chi^2(2, N = 573) = 1.41, p = .495$.

One limitation of this review of reviews was that there were no estimates of interrater reliability for the variables that were coded. However, that limitation is mitigated by the fact that the coding variables were not of a subjective nature. In Table 4, I listed all of the previous categories that had been used and made explicit how they related to the metacategory variable. Arriving at the proportions for the metacategories was then simply a matter of summing the number of articles that belonged to each of the subcategories in the metacategory.

In summary, I found that most of the research in educational technology had been quantitative, some of it qualitative, and a small percentage of it mixed methods. The percentage of empirical papers that dealt with human participants was much higher in education research proper than in educational technology. However, the relative proportions of quantitative, qualitative, and mixed-methods articles in educational technology and education research proper forums were about the same.

Methodological Reviews in Computer Science Proper, Software Engineering, and Information Systems

Although ancillary to computer science education, there are three seminal methodological reviews of the computer science literature proper that are worth mentioning and that may help put the results of this book into context. Those reviews are Glass, Ramesh, and Vessey (2004); Tichy, Luckowicz, Prechelt, and Heinz (1995); and Zelkowitz and Wallace (1997).

In "An Analysis of Research in Computing Disciplines," Glass et al. (2004) reviewed 1,485 articles from a selection of journals in the fields of computer science, software engineering, and information systems. They classified each article by topic, research approach, research method, reference discipline, and level of analysis. Some findings from the Glass et al. review that might be relevant to the current review are quoted below:

CS [computer science] research methods consisted predominantly of mathematically based Conceptual Analysis (73%). SE [software engineering] used Conceptual Analysis that is not mathematically based (44%) with Concept Implementation also representing a significant research method at 17%. IS [information systems] research used predominantly five types of research methods, the most notable being Field Study (27%), Laboratory Experiment (Human), Conceptual Analysis (15%), and Case Study (13%). (p. 92)

In "Experimental Evaluation in Computer Science: A Quantitative Study," Tichy et al. (1995) did a methodological review of 400 articles from complete volumes of several refereed computer science journals, a conference, and 50 titles drawn at random from all articles published by ACM [The Association for Computing Machinery] in 1993. The journals of *Optical Engineering (OE)* and *Neural Computation (NC)* were used for comparison. (p. 9)

They classified each article according to several attributes, such as whether it was an empirical work or not. The major findings from the Tichy et al. review are quoted below:

> Of the papers in the random sample that would require experimental validation, 40% have none at all. In journals related to software engineering, this fraction is 50%. In comparison, the fraction of papers in OE [a journal called *Optical Engineering*] and NC [a journal called *Neural Computing*] is only 15% and 12%, respectively. Conversely, the fraction of papers that devote one fifth or more of their space to experimental validation is almost 70% for *OE* and *NC*, while it is a mere 30% for the computer science (CS) random sample and 20% for software engineering. The low ratio of validated results appears to be a serious weakness in computer science research. This weakness should be rectified for the long-term health of the field. (p. 9)

Zelkowitz and Wallace (1997), in "Experimental Validation in Software Engineering," reviewed over 600 papers from the software engineering literature and 100 articles from other fields as a basis for comparison. As in the other reviews, they classified the articles into methodological categories. Some of their findings that are relevant to the current review are presented below:

> We observed that 20% of the papers in the journal *IEEE Transactions on Software Engineering* have no validation component (either experimental or theoretical). This number is comparable to the 15 to 20% observed in other scientific disciplines. However, about one-third of the software engineering papers had a weak form of experimentation (assertions) where the comparable figure for other fields was more like 5 to 10%. (p. 742)

Several things need to be noted about these reviews before using them as a basis for comparison with computer science education research. First, it is difficult, if not impossible, to synthesize the results of these reviews because each uses a different categorization system. Second, the relevance of these reviews to the field of computer science education is questionable; these reviews only apply to computer science education research in as much as computer science education research was a part of the samples of the computer science, software engineering, and information systems literature that were reviewed. Finally, some question the validity of these reviews. For example, Tedre (2006) argued that the Glass et al. (2004) study "may not adequately describe what actually happens in computer science" (p. 349), that the granularity of the categories in Glass et al.'s study is overly coarse, and that "the choice of mainstream journals may have biased the sample of articles towards mainstream research so that alternative methods may get lesser attention" (p. 349).

The Scope and Quality of the Previous Methodological
Reviews of Computer Science Education Research

The argument that has been developed thus far is that methodological reviews have been used successfully to improve the methodological practices of researchers in a variety of behavioral research fields, and the conditions appear met for methodological reviews to also help improve the emerging methodological practices of computer science education researchers. Although there have been several methodological reviews of research on computer science education, I will demonstrate in the following section that those methodological reviews are limited either in their breadth, depth, or reliability.

To identify all the past methodological reviews of computer science education, six searches of the Internet; the ACM Digital Library; and Academic Premier, Computer Source, ERIC, Psychology and Behavioral Science Collections, and PyscINFO (via Ebsco Host) were conducted on November 29, 2005 using the keyword combinations: *"computer science education research," "methodological review,"* and *"computer science education research," "meta-analysis."* Another six searches on January 20, 2006 were conducted using the same databases but using the keyword combinations: *"computer science education research," "systematic review,"* and *"computer science education research," "research synthesis."* The summary, title, or abstract of each record was read to determine if it would lead to a review of the research methods in computer science education.

In addition to the electronic searches, the table of contents of (a) the *Koli Calling Proceedings* (2001-2005), (b) the *ICER Proceedings 2005*, (c) *Computer Science Education* (volumes 8-15), and (d) the *Journal of Computer Science Education Online* (the volumes published between 2001-2005) were searched. Also, a pearl-building approach was taken to identify other reviews from the reference sections of the reviews, including meta-analyses, found from the searches described above. Meta-analyses, or other reviews that emphasized research outcomes rather than methods, were excluded from this review of computer science education methodological reviews. The term *meta-analysis* was included as a search term because sometimes methodological reviews are mislabeled as meta-analyses, as was the case with Valentine's article (2004). Table 7 shows the number of records that resulted from each search.

Based on the search procedure mentioned above, I found that three methodological reviews of computer science research (or evaluation) had been conducted since computer science education research began in the early 1970s. (One review that should be acknowledged, but was not classified as a methodological review is Kinnunen [n.d.]. In that review, Kinnunen examined the *subject matter*

Table 7. *Description of the Electronic Search for Previous Methodological Reviews*

Search	Term	Database	Records
1	"computer science education research" "methodological review"	Internet (Google)	0
2	"computer science education research" "meta-analysis"	Internet (Google)	10
3	"computer science education research" "systematic review"	Internet (Google)	3
4	"computer science education research" "research synthesis"	Internet (Google)	1
5	"computer science education research" "methodological review"	ACM library	27
6	"computer science education research" "meta-analysis"	ACM library	315
7	"computer science education research" "systematic review"	ACM library	33
8	"computer science education research" "research synthesis"	ACM library	21
9	"computer science education research" "methodological review"	Ebsco Host	0
10	"computer science education research" "meta-analysis"	Ebsco Host	0
11	"computer science education research" "systematic review"	Ebsco Host	0
12	"computer science education research" "research synthesis"	Ebsco Host	0

of the articles published in SIGCSE Bulletin. Those three reviews (Randolph, 2005; Randolph, Bednarik, & Myller 2005; Valentine, 2004) were already presented in detail in the section entitled *"Computer Science Education Research is an Emerging Field,"* so they will not be presented again here. I will, however, describe their samples and map the areas of computer science education research that have been covered. Before that, however, I will explain my assumption of what the population of computer science education research reports consist of.

In this investigation, I was interested in making a generalization to the entirety of recent research published in the major computer science education research forums. I operationalized this as the full papers published from 2000 to 2005 as the June and December issues of *SIGCSE Bulletin* [hereafter *Bulletin*], a computer science education journal; *Computer Science Education* [hereafter *CSE*], a computer science education research journal; the *Journal of Computer Science Education Online*, [hereafter *JCSE*], a little-known computer science education journal; the *Proceedings of the SIGCSE Technical Symposium* [hereafter *SIGCSE*]; The *Proceedings of the Innovation and Technology in Computer Science Education Conference* [hereafter *ITiCSE*]; the *Koli Calling: Finnish/Baltic Sea Conference on Computer Science Education* [hereafter *Koli*], the *Proceedings of the Australasian Computing Education Conference* [hereafter ACE], and the *International Computer Science Education Research Workshop* [hereafter *ICER*]. (The fall and spring issues of *Bulletin* are the *SIGCSE* and *ITiCSE* proceedings.) I included "full papers," but excluded poster summaries, demo summaries, editorials, conference reviews, book reviews, forewords, introductions, and prologues in the sampling frame. The three previous methodological reviews of computer science education research (Randolph, 2005; Randolph, Bednarik, & Myller, 2005; Valentine, 2004) only cover a very small part of the population operationalized above. Additionally, the review that is most representative of the population of computer science education research articles (Valentine) has serious methodological flaws.

In the Randolph, Bednarik, and Myller (2005) methodological review, a census of the full papers published in the *Proceedings of the Koli Calling Conference* from 2001 to 2004 was reviewed. Although a census was conducted, the articles in the *Proceedings of the Koli Calling Conference* made up only a small, marginal part of the population of recent computer science education research articles. For example, the articles published in the *Proceedings of the Koli Calling Conference* from 2001 to 2005 only accounted for 7% of the population specified above. Also, the *Koli Calling Conference* is a regional conference (Finnish/Baltic) and, therefore, its proceedings are not representative of the population of computer science education research articles as a whole. For example, about 90% of the papers in the Randolph et al. review were of Finnish origin.

The Randolph (2005) methodological review focused on a subset of the grey literature on computer science education—reports of evaluations of computer science education programs.

(Almost all of the program evaluation reports included in the review of program evaluation reports were published on the Internet or in the ERIC database.) In the methodological review section of the Randolph review, 29 program evaluation reports were analyzed. Of those 29, only two of the reviewed reports had been summarized in one of the forums included in my operationalization of the computer science education research population. Thus, the population of the Randolph review is almost entirely different than the population of this report.

The Valentine (2004) methodological review included 444 articles that dealt with the first year of computer science education courses and were published in the *SIGCSE Technical Symposium Proceedings* from 1984 to 2003. Valentine reviewed a large number of articles, but he sampled them from only one forum for publishing computer science education research and excluded articles that did not deal with first-year computer science courses. In addition to the potentially low generalizability of Valentine's sample, the quality of the Valentine review is questionable. First, Valentine only coded one variable for each article—he simply classified the articles into one of six categories: *Marco Polo, Tools, Experimental, Nifty, Philosophy,* and *John Henry.* The *experimental* category—operationalized as "any attempt at assessing the 'treatment' with some scientific analysis" (Valentine, p. 256)—is so broad that it is not useful as a basis for recommending improvements in practice. Second, Valentine coded all of the articles himself without any measure of interrater agreement.

In conclusion, the three previous methodological reviews either lacked breadth, depth, or reliability. Randolph, Bednarik, and Myller (2005), Randolph (2005), and, to a lesser extent, Valentine (2004) do not represent the population of published computer science education research. What is more, the Valentine review, which has the greatest number of reviewed articles, has questionable reliability. Also, Valentine only coded the articles in terms of one somewhat light-hearted variable. Given that fact, it is difficult to say with certainty what the methodological practices in computer science education research are and, consequently, it is also difficult to have a convincing basis to suggest improvements in practice.

Purpose and Research Questions

Because the past methodological reviews of computer science education research had limitations either in terms of their generalizability or reliability, I conducted a replicable, reliable, methodological review of a representative sample of the research published in the major computer science education forums over the last 6 years. This research (a) provides significantly more-representative coverage of the field of computer science education than any of the previous reviews, (b) covers articles with more analytical depth (with a more-refined coding sheet) than any of the previous reviews, and (c) with a greater amount of reliability and replicability than any of the other previous reviews. In short, this book simultaneously extends the breadth, depth, and reliability of the previous reviews.

The purpose of this methodological review was to have a valid and convincing basis on which to make recommendations for the improvement of computer science education research and to promote informed dialogue about its practice. If my recommendations are heeded and dialogue increases, computer science education is expected to improve and, consequently, help meet the social and economic needs of a technologically oriented future.

To have a valid basis to recommend improvements of computer science education research methodology, I answered the primary research question: *What are the methodological properties of research reported in articles in major computer science education research forums from the years 2000-2005?* The primary research question can be broken down into several subquestions, which are listed below:

1. What was the proportion of articles that reported research on human participants?
2. Of the articles that did not report research on human participants, what types of articles were being published and in what proportions?
3. Of the articles that did report research on human participants, what proportion provided only anecdotal evidence for their claims?
4. Of the articles that did report research on human participants, what types of methods were used and in what proportions?
5. Of the articles that did report research on human participants, what measures were used, in what proportions, and was psychometric information reported?
6. Of the articles that did report research on human participants, what were the types of independent, dependent, mediating, and moderating factors that were examined and in what proportions?
7. Of the articles that used experimental/quasi-experimental methods, what typesof designs were used and in what proportions? Also, were participants randomly assigned or selected?
8. Of the articles that reported quantitative results, what kinds of statistical practices were used and in what proportions?

9. Of the articles that did report research on human participants, what were the characteristics of the articles' structures?

Based on the previous methodological reviews of computer science education research, I made predictions for eight of the nine subquestions above. This research tested those predictions on a random sample of the entire population of articles or conference papers published in major computer science education research forums. The predictions are listed below; the citations refer to the source(s) from which the prediction was made.

1. Between 60% and 80% of computer science education research papers will not report research on human participants (Randolph, 2005; Randolph, Bednarik, & Myller, 2005).
2. Of the papers that do not report research on human subjects, the majority (about 60%) will be purely program (intervention) descriptions (Randolph, Bednarik, & Myller, 2005; Valentine, 2004).
3. Of the articles that do report on human participants, about 15% will report only anecdotal evidence for their claims (Randolph, Bednarik, & Myller, 2005).
4. Of the articles that report research on human participants, articles will most frequently be reports of experiments/quasi-experiments or exploratory descriptions (e.g., survey research), as opposed to correlational studies, explanatory descriptive studies (e.g., qualitative types of research), causal-comparative studies, or classification studies; (Randolph, 2005; Randolph, Bednarik, & Myller, 2005).
5. Of the articles that do report research on human participants, questionnaires, grades, and log files will be the most frequently used types of measures. None (or very few) of the measures will have psychometric information reported (Randolph, 2005; Randolph, Bednarik, & Myller, 2005).
6. Of the articles that do report research on human participants, the most frequent type of independent variable will be student instruction, the most frequent dependent variable will be stakeholder attitudes, and the most frequent moderating variable will be gender (Randolph, 2005; Randolph, Bednarik, & Myller, 2005).
7. Of the articles that report experiments or quasi-experiments, the one-group posttest-only design and posttest-only with controls design will be the most frequently used types of experimental designs. Instances of random selection or random assignment will be rare (Randolph, 2005; Randolph, Bednarik, & Myller, 2005).

8. Of the articles that report research on human participants, about 50% of the reports will be missing a literature review section. The vast majority will not have explicitly stated research questions. (Randolph, Bednarik, & Myller, 2005).

In addition to answering the primary research question—What are the methodological characteristics of the computer science education research published in major forums between 2000 and 2005? —I conducted 15 planned contrasts to identify islands of practice. In the contrasts, there were three comparison variables—(a) type of publication forum: journal or conference proceedings, (b) year, and (c) region of first author's institutional affiliation—crossed by five dependent variables: (a) frequency of articles in which only anecdotal evidence was reported; (b) frequency of articles that reported on experimental or quasi-experimental investigations; (c) frequency of articles that reported on explanatory descriptive investigations; (d) frequency of experimental or quasi-experimental articles that used a one-group posttest-only research design exclusively; and (5) the frequency of articles in which attitudes were the only dependent variable measured.

The 15 planned contrasts answered the following three secondary research questions:

1. Is there an association between type of publication (whether articles are published in conferences or in journals) and frequency of articles providing only anecdotal evidence, frequency of articles using experimental/quasi-experimental research methods, frequency of articles using explanatory descriptive research methods, frequency of articles in which the one-group posttest-only design was exclusively used, and frequency of articles in which attitudes were the sole dependent variable?

2. Is there a yearly trend (from 2000-2005) in terms of the frequency of articles providing only anecdotal evidence, frequency of articles using experimental/quasi-experimental research methods, frequency of articles using explanatory descriptive research methods, frequency of articles in which the one-group posttest-only design was exclusively used, and frequency of articles in which attitudes were the sole dependent variable?

3. Is there an association between the region of the first author's institutional affiliation and frequency of articles providing only anecdotal evidence, frequency of articles using experimental/quasi-experimental research methods, frequency of articles using

explanatory descriptive research methods, frequency of articles in which one-group posttest-only designs were exclusively used, and frequency of articles in which attitudes were the sole dependent variable?

Note that the primary and secondary questions that were asked here are basically the same questions that were asked in methodological reviews in a closely related field—educational technology (see Table 2). Also, the question regarding the statistical practices of computer science education researchers (i.e., Subquestion 8 of the primary research question) was aligned with the main questions that were asked in the methodological reviews that supported the APA Task Force on Statistical Inference's recommendations.

In addition to investigating islands of practice within the field of computer science education, I also investigated islands of practice between the related fields of computer science education, educational technology, and education research proper. My research question in this area follows: How do the proportions of quantitative, qualitative, and mixed methods articles in computer science education compare to those proportions in the fields of educational technology and education research proper?

Tedre (2006) explained that computer science is a field that is comprised, mainly, of three traditions: a formalist tradition, an engineering tradition, and an empirical tradition. I predicted that this engineering tradition would make itself most evident in computer science education research, and to a lesser degree in education technology (because it also consists of an engineering component; Ely [1999], one of the key figures in education technology, calls it a "physical sciences component"), and reflected least in education research proper. Here I assume that the number of papers that are program descriptions (i.e., papers that do not empirically deal with human participants) is an indicator of the degree of the engineering and formalist traditions in the fields of computer science education, educational technology, and education research proper.

Specifically, if my prediction is correct then I would expect to find that computer science education research forums have the highest proportions of program descriptions (engineering) articles (e.g., *I built this thing to these specifications* types of articles), educational technology forums would have the second highest proportions of program descriptions articles, and that education proper forums would have the lowest proportions of program descriptions article, but would have the highest proportion of empirical articles dealing with human participants.

Biases

My background is in behavioral science research (particularly quantitative education-research and program evaluation); therefore, I brought the biases of a quantitatively trained behavioral scientist into this investigation. It is my belief that when one does education-related research on human participants the conventions, standards, and practices of behavioral research should apply; therefore, I approached this methodological review from a behavioral science perspective. Nevertheless, I realize that computer science education and computer science education research is a maturing, multidisciplinary field, and I acknowledge that the behavioral science perspective is just one of many valid perspectives that one can take in analyzing computer science education research.

Chapter 2

Review Methods

Neuendorf's (2002) *Integrative Model of Content Analysis* was used as the model for carrying out the proposed methodological review. Neuendorf's model consists of the following steps: (a) developing a theory and rationale, (b) conceptualizing variables, (c) operationalizing measures, (d) developing a coding form and coding book, (e) sampling, (f) training and determining pilot reliabilities, (g) coding, (h) calculating final reliabilities, and (i) analyzing and reporting data.

In the following subsections, I describe how I conducted each of the steps of Neuendorf's model. Because the rationale (the first step in Neuendorf's model) was described earlier, I do not discuss it below.

Conceptualizing Variables, Operationalizing Measures, and Developing a Coding Form

Because this methodological review was the sixth in a series of methodological reviews I had conducted (see Randolph et al., 2004; Randolph, 2005; Randolph, in press; Randolph, Bednarik, & Myller, 2005; Randolph, Bednarik, Silander, et al., 2005; and Randolph & Hartikainen, 2005), most of the variables had already been conceptualized, measures had been operationalized, and coding forms and coding books had been created in previous reviews. A list of the articles that were sampled are included in Appendix A. The coding form and coding book that I used for this methodological review are included as Appendices B and C, respectively.

Sampling

A proportional stratified random sample of 352 articles, published between the years 2000 and 2005, were drawn, without replacement, from the eight major peer-reviewed computer science education publications. (That sample size, 352, out of a finite population of 1,306 was determined a priori, through the *Sample Planning Wizard* [2005] and confirmed through resampling, to be large enough to achieve a +/- 5% margin of error with a 95% level of statistical confidence if I were to treat all variables, and levels of variables, as dichotomous, in the most conservative case—where

---

33

p and $q = .5$. This power estimate refers to the aggregate sample, not to subsamples.) The sample was stratified according to year and source of publication. Table 8, the sampling frame, shows the number of papers (by year and publication) that existed in the population as I operationalized it. Table 5 shows the number of articles that were randomly sampled (by year and publication source) from each cell of the sampling frame presented in Table 9. The articles that were included in this sample are listed in Appendix A.

The population was operationalized in such a way that it was a construct of what typically is accepted as mainstream computer science education research. The population did not include the marginal, grey areas of the literature such as unpublished reports, program evaluation reports, or other nonpeer-reviewed publications because I was not interested in the research practices

Table 8. *Sampling Frame*

Year/forum	2000	2001	2002	2003	2004	2005	Total
Bulletin	31	21	21	40	36	38	187
CSE	17	17	17	17	17	15	100
JCSE	0	2	7	5	2	2	18
KOLI	0	14	10	13	21	25	83
SIGCSE	78	78	74	75	02	104	501
ITICSE	45	44	42	41	46	68	286
ICER	0	0	0	0	0	16	16
ACE	0	0	0	34	48	33	115
Total	171	176	171	225	262	301	1306

Table 9. *Number of Articles Sampled from Each Forum and Year*

Year/forum	2000	2001	2002	2003	2004	2005	Total
Bulletin	8	6	6	11	10	10	51
CSE	5	5	5	5	5	4	29
JCSE	0	0	2	1	0	0	3
KOLI	0	4	3	3	6	7	23
SIGCSE	21	21	20	20	25	28	135
ITICSE	12	12	11	11	12	13	76
ICER	0	0	0	0	0	4	4
ACE	0	0	0	9	13	9	31
Total	46	48	47	60	71	80	352

reported in the entirety of computer science education research. Rather, I was interested in research practices reported in current, peer-reviewed, mainstream computer science education research forums.

In general, nonpeer-reviewed articles or poster-summary papers (i.e., papers two or fewer pages in length) were not included in the sampling frame. In *Bulletin*, only the peer-reviewed articles were included; featured columns, invited columns, and working group reports were excluded in the sampling frame of Table 8. In *CSE* and *JCSE*, editorials and introductions were excluded. In the *SIGCSE, ITICSE, ACE,* and *ICER* forums, only full peer-reviewed papers at least three pages in length were included; panel sessions and short papers (i.e., papers two pages or less in length) were excluded. In *Koli,* research and discussion papers were included; demo and poster papers were excluded.

Training and Determining Pilot Reliabilities

In this methodological review, an interrater reliability reviewer, who had participated in previous methodological reviews, was trained in the coding book and coding sheet, which are included as Appendices B and C. The interrater reliability reviewer, Roman Bednarik, was a PhD student in computer science at the University of Joensuu. He was chosen because he had significant knowledge of computer science, computer science education, and quantitative research methodology and because he had participated in previous methodological reviews of computer science education or educational technology research. (Randolph, Bednarik, & Myller, 2005; Randolph, Bednarik, Silander, et al., 2005). AlthouGH his knowledge and previous experience in collaborating on methodological reviews meant that he required less coder training than if a different coder had been chosen, it also meant that he was aware of my hypotheses about computer science education research.

Initially the interrater reliability reviewer and I read through the coding book and coding sheet together and discussed any questions that he had about the coding book or coding sheet. When inconsistencies or ambiguities in the coding book or coding sheet were found in the initial training session, the coding book or coding sheet was modified to remedy those inconsistencies or ambiguities. Then the interrater reliability reviewer was given a revised version of the coding book and coding sheet and was asked to independently code a purposive pilot sample of 10 computer science education research articles, which were not the same articles that were included in the final reliability subsample. The purposive sample consisted of articles that I deemed to be representative of the different types of research methods that were to be measured, articles that were anecdotal only, and articles that did not deal with human participants. I, the primary coder, also coded those 10 articles. After both of us had coded the 10 articles we came together to compare our codes and to discuss the inconsistencies or unclear directions in the coding book and coding sheet. When we had disagreements about article codes, we would try to determine the cause of the disagreement and I would modify the coding book if it were the cause of the disagreement. After pilot testing and subsequent improvement of the coding book and the coding, the final reliability subsample was coded (see the section entitled *Calculating Final Reliabilities*).

Since many of the variables in the coding book were the same as in previous reviews (specifically, Randolph, 2005; Randolph, Bednarik, & Myller, 2005; Randolph, Bednarik, Silander, et al., 2005), many of the pilot reliabilities had already been estimated. The variables that had been used in previous reviews and already had estimates of interrater reliabilities were methodology category; type of article, if not dealing with human participants; whether an experimental or quasi-experimental design was used; type of selection and assignment; psychometric information provided; type of experimental or quasi-experiment; structure of the paper (i.e., report elements); measures; independent variables; dependent variables; and moderating or mediating variables. (See Randolph, 2005; Randolph, Bednarik, & Myller, 2005; and Randolph, Bednarik, Silander, et al., 2005 for previous estimates and discussions of interrater reliabilities for these variables.) In general, all of the reliabilities for these variables were, or eventually became, acceptable or the source of the unreliability had been identified and had been remedied in the current coding book (see Randolph, Bednarik, & Myller). The only set of variables whose reliabilities had not been pilot tested in previous methodological reviews dealt with statistical practices or were demographic variables. Reliabilities for the demographic characteristics, such as name of the first author, were not estimated since they were objective facts.

Coding

Appendices B and C, which are the coding sheet and coding book, provide detailed information on the coding variables, their origin, and the coding procedure. Because the complete coding sheet and coding book are included as appendices, I will only summarize them here.

Articles were coded in terms of demographic characteristics, type of article, type of methodology used, type of research design used, independent variables examined, dependent and mediating measures examined, moderating variables examined, measures used, and statistical practices. In the rest of this section I describe the variables in the coding book and their origin and history.

The first set of variables, demographic characteristics, consisted of the following variables:

- The case number,
- The case number category (the first two digits of the case number),
- Whether it was a case used for final reliability estimates,
- The name of the reviewer,
- The forum from which the article came,
- The type of forum from which the article came (i.e., a journal or conference proceedings),
- The year the article was published,
- The volume number where the article was published,
- The issue in which the article was published,
- The page number on which the article began,
- The number of pages,
- The region of the first author's affiliation,
- The university affiliation of the first author,
- The number of authors, and
- The last name and first initials of the first author.

The variables in the second set, type of article, are listed below:

- Kinnunen's categories;
- Valentine's categories;
- Whether the article dealt with human participants;
- If the article did not deal with human participants, what type of article it was; andIf the article did deal with human participants, whether it presented only anecdotal evidence.

The Kinnunen's categories variable was derived from Kinnunen (n.d.). The Valentine's category variable was derived from Valentine (2004). The rest of the variables in this section were originally derived from an emergent coding technique in Randolph, Bednarik, Silander, and colleagues (2005) and then refined and used in Randolph, Bednarik, and Myller (2005) before being refined again and used in the current coding book.

The third set of variables, report structure, originated in the *Parts of a Manuscript* section of the *Publication Manual of the American Psychological Association* (2001). The exceptions are the grade level and curriculum year variables, which were suggested by committee members during the proposal defense of this book. The report structure variables are listed below:

- Type of abstract,
- Introduction to problem present,
- Literature review present,
- Purpose/rational present,
- Research questions/hypotheses present,
- Adequate information on participants present,
- Grade level of students,
- Curriculum level taught,
- Information about settings present,
- Information about instruments present,
- Information about procedure present, and
- Information about results and discussion present.

The fourth set of variables, methodology type, was developed from Gall, Borg, and Gall (1996) and from the *Publication Manual of the American Psychological Association* (APA, 2001). The explanatory descriptive and exploratory descriptive labels came from Yin (1988). The descriptions of these variables in the coding book evolved into their current form though Randolph (2005, in press), Randolph, Bednarik, and Myller (2005), and Randolph, Bednarik, Silander, and colleagues. (2005). The assignment variable originated from Shadish, Cook, and Campbell (2002). The methodology type variables are listed below:

- Whether the article reported on an experimental or quasi-experimental investigation or not,
- Whether the article reported on an explanatory descriptive investigation or not,
- Whether the article reported on an exploratory descriptive investigation or not,
- Whether the article reported on a correlational investigation or not,
- Whether the article reported on a causal-comparative investigation or not,
- If there was not enough information to determine what type of method was used, and
- The type of selection used.

The fifth set of variables, experimental research designs, relate to the articles that reported on an experimental or quasi-experimental investigation. If experimental or quasi-experimental investigations were reported, the type of experimental or quasi-experimental design was noted. These research design variables were derived from Shadish, Cook, and Campbell (2002) and from the *Publication Manual of the American Psychological Association* (APA, 2001). These variables had been previously pilot tested in Randolph (2005; in press), Randolph, Bednarik, and Myller (2005), and Randolph, Bednarik, Silander, and colleagues (2005), except for the multiple factor variable, which had not been previously pilot tested. The experimental research design variables are listed below:

- If there was enough information to determine what experimental design had been used if one had been used,
- If the researchers used a one-group posttest-only design,
- If the researchers used a posttest with controls design,
- If the researchers used a pre/posttest without controls design,
- If the researchers used a pre/posttest with controls design,
- If the researchers conducted a repeated measures investigation,
- If the researchers used a design that involved multiple factors, and
- If the researchers used a single-case design.

The sixth set of variables dealt with the type of independent variables that were reported. These variables were derived through an emergent coding technique from Randolph (2005) and Randolph, Bednarik, and Myller (2005). The binary independent variables listed in the coding book for this set of variables are listed below:

- Student instruction,
- Teacher instruction,
- Computer science fair or contest,
- Mentoring,
- Listening to computer science speakers,
- Computer science fields, and
- Other types of interventions (open variable).

The seventh set of variables in the coding book dealt with the types of dependent variables that were measured. These variables were based on codes that emerged from Randolph (2005) and Randolph, Bednarik, and Myller (2005). The variables in this set are listed below:

- Attitudes (including self/reports of learning),
- Attendance,
- Achievement in core courses,
- Achievement in computer science,
- Teaching practices,
- Students' intentions for the future,
- Program implementation,
- Costs,
- Socialization,
- Computer use, or
- Other types of dependent variables (open variable).

The eighth set of variables dealt with the types of measures that computer science educators used. These measurement variables were derived from codes that emerged in Randolph (2005) and Randolph, Bednarik, and Myller (2005). Those binary measurement variables are listed below:

- Grades,

- Student diaries,

- Questionnaires,

- Log files,

- Teacher- or researcher-made tests,

- Interviews,

- Direct observation,

- Standardized tests,

- Student work,

- Focus groups,

- Existing records, or

- Other types of measures (open variables).

Additionally whether any sort of psychometric information was provided for the variables involving questionnaires, teacher- or researcher-made tests, direct observation, or standardized tests.

The ninth set of variables involved mediating or moderating variables. In the coding book this set of variables are called *Factors (Non-manipulatable variables)*. This set of variables was based on codes that emerged from Randolph (2005) and Randolph, Bednarik, and Myller (2005). Those variables are listed below:

- Gender,

- Aptitude,

- Race/ethnic origin,

- Nationality,

- Disability,

- Socioeconomic status, and

- Other types of dependent variables (open variables).

The tenth and final set of variables involved statistical practices. The statistical practices variables dealt mainly with how inferential statistics and effect sizes were used and reported. Particular emphasis was placed on whether informationally adequate statistics were provided for a certain type of analysis. What was considered to be an informationally adequate set of statistics is discussed in detail in the coding book. These variables were based on the guidelines in *Informationally Adequate Statistics* section of the *Publication Manual of the American Psychological Association* (APA, 2001). The variables in that set are listed below:

- Whether quantitative results were reported,
- Whether inferential statistics were reported,
- Whether parametric tests were conducted and an informationally adequate set of statistics were reported for them,
- Whether multivariate analyses were conducted and an informationally adequate set of statistics was reported for them,
- Whether correlational analyses were conducted and an informationally adequate set of statistics was reported for them,
- Whether parametric analyses were conducted and an informationally adequate set of statistics was reported for them, and
- Whether analyses for small samples were conducted and an informationally adequate set of statistics was reported.

In addition to the variables related to inferential practices, there was also a set of variables about what types of effect sizes were reported. Those variables are listed below:

- Whether an effect size was reported,
- Whether a raw difference effect size was reported,
- Whether a standardized mean difference effect size was reported,
- Whether a correlational effect size was reported,
- Whether odds ratios were reported,
- Whether odds were reported, and
- Whether some other type of effect size other than the ones above were reported (an open

variable).

In terms of the coding procedure, the primary coder (the author of this text) used the coding sheet and coding book to code a stratified random sample of 352 articles. A subsample of 53 articles was selected randomly from those 352 articles and electronic files of those 53 articles was given to the interrater reliability coder, who also used the coding sheet and coding book to code those 53 articles. The primary coder and interrater reliability coder did not converse about the coding process while the coding was being done. After the coding was completed the primary coder merged the two sets of codes for the subsample and calculated interrater reliability estimates. When there were disagreements about the coding categories, the primary coder's judgment took precedent. Variable-by-variable instructions for the coding procedure are given in the coding book.

Calculating Final Reliabilities

According to Neuendorf (2002), a reliability subsample of between 50 and 300 units is appropriate for estimating levels of interrater agreement. In this case, a simple random reliability subsample of 53 articles was drawn from the sample of 352 articles. Those 53 articles were coded independently by the interrater reliability reviewer so that interrater reliabilities could be estimated.

Because the marginal amounts of each level of variables to be coded were not fixed, Brennan and Prediger's (1981) free-marginal kappa was used as the statistic of interrater agreement. (By fixed, I mean that there was not a fixed number of articles that must be assigned to given categories. The marginal distributions were free. See Brennan & Prediger, 1981.) Values of kappa lower than .4 were considered to be unacceptable, values between .4 and .6 were considered to be poor, values between and including .6 and .8 were considered to be fair, and values above .8 were considered to be good reliabilities. Confidence intervals around kappa were found through resampling.

Data Analysis

To answer the primary research question, I reported frequencies for each of the multinomial variables or groups of binominal variables. Confidence intervals (95%) for each binary variable or multinomial category were calculated through resampling (see Good, 2001; Simon, 1997), "an alternative inductive approach to significance testing, now becoming more popular in part because of the complexity and difficulty of applying traditional significance tests to complex samples" (Garson, 2006, n.p). The *Resampling Stats* language (1999) was used with the Grosberg's (n.d.) resampling program.

To answer the research questions that involved finding islands of practice, I took two approaches. In the first approach, I cross tabulated the data for the 15 planned contrasts, examined the adjusted residuals, and, for categorical variables calculated χ^2 (see Agresti, 1996) and found its probability through resampling. For ordinal variables, such as year, I calculated M^2 (see Agresti) and found its probability through resampling. The resampling codes for calculating χ^2 and M^2 from a proportionally stratified random sample can be found in Appendix F. In the second approach, I used logistic regression to determine the unique effect of the three predictor variables (i.e., forum type, region of first author's affiliation, and year) on the five binary outcome variables (i.e., anecdotal-only paper, experimental/quasi-experimental paper, explanatory descriptive paper, attitudes-only paper, or one-group posttest-only paper) and to determine if there were interactions between the variables.

To carry out the logistic regression, with *SPSS* 11.0, I followed the method described in Agresti (1996). First, I found the best fitting logistic regression model for each outcome variable by starting with the most complex model, which had the main effects, all two-way interactions, and the one three-way interaction (i.e., I+R+Y+F+R *Y+R*F+Y*F+R*Y*F; where I = intercept, R = region of first author's affiliation [a categorical variable], F = forum type [journal or conference proceeding] [a categorical variable], and Y = year), and then reducing the complexity of the model until the point when the less-complex model would raise the difference in the deviances between the two models to a statistically significant level. To determine if a less-complex model was as good fitting as the more-complex model, I took the absolute value of the difference in the -2 Log Likehood [hereafter *deviance*] and degrees of freedom between each model and used the χ^2 distribution to determine if there was a statistically significant increase in the deviance. For example, if a full model had a deviance of 286.84 and 11 degrees of freedom and the model without the three-way interaction had a deviance of 289.93 and 9 degrees of freedom, the difference between models would be 1.09 in deviance and 2 degrees of freedom. The χ^2 probability associated with those values is .58. Because the difference was not statistically significant, I concluded that the less-complex model was, more or less, as well fitting (i.e., it had about an equal amount of deviance) as the more-complex model. I repeated this process until I found the least complex model that had a deviance about equal to the deviance of the next most complex model. If the best fitting model

44

was overspecified (i.e., if the continuous, year variable was not in the best-fitting model), I included the year variable nonetheless to fix the overspecification problem and ran both analyses, with and without the continuous variable.

I relied on several methods to determine the overall fit of the model to the data. I used *SPSS*'s Omnibus Test of Model Coefficents (i.e., χ^2 of the difference of the selected model and the model with only a constant), which should be statistically significant if the chosen model is better than the model with only a constant (Agresti, 1996). I also used *SPSS*'s version of the Hosmer and Lemeshow test, which breaks the data set into deciles and computes the deviation between observed and predicted values. If the model fits appropriately, the Hosmer and Lemeshow test should not be statistically significant (Agresti). Also, I created scatterplots of the expected and observed probabilities. If visual inspection of the plots showed that there were outliers, I ran regression analyses with and without the outliers removed. Finally, I also examined the regression coefficients to determine if the model seemed to fit the data. For example, if there were exponentiated coefficients (odds ratios) in the thousands, I would use a different model or group the data in a different way. To illustrate, in some cases I found that I had to group some of the regions together to get enough cases in a category for the regression coefficients to make sense.

C h a p t e r 3

Review Results

To eliminate a significant rounding error when automating the resampling analysis, I had to slightly overestimate the population size so that the ratio of population- to-sample was an integer. Without this overestimation, the rounding error caused the resampled parameter proportions to differ significantly from the sample proportions—sometimes the two proportions would differ by as much as 5%. The actual population to sample ratio was 3.71/1 (or 1,306/352), but in my analysis I rounded the ratio's numerator to the next nearest integer, 4. In terms of my analyses, my estimate of the finite population was 1,408 (4*352) instead of 1,306. The statistical consequences are that overestimating the population will lead to slightly conservative results (Kalton, 1983); however, in this case the differences between using a population of 1,306 and 1,408 were negligible. Using Formula 11 of Kalton (p. 21) to manually estimate the confidence intervals around a proportion, in this case around the proportion of *human participants* variable, the proportion of the standard error when using a population of 1,306 (1.84) to the standard error when using a population of 1,408 (1.86) was 0.99. Or, from a different viewpoint, the length of confidence intervals when using a population size of 1,306 was 7.30 percentage units long and when using a population size of 1408 the length of the confidence interval was 7.21 percentage units long—a 9/100% difference in the length of the confidence intervals.

According to Agresti (1996), regrouping data sometimes is necessary when working with categorical data. In this case it was necessary to group the regions of first author's affiliations together in order for certain statistical analyses, such as logistic regression, to work. For example, in some of the logistic regression equations I had to group the regional categories with the fewest cases into one group, because they had so few observations at fine levels of analysis. Specifically, I sometimes grouped some of the region of first author's affiliation categories—Africa, Asia-Pacific/Eurasia, and Middle East—into one category that I called *Asia-Pacific/Eurasia et al.* My rationale for this grouping is that although I could no longer make distinctions between African,

Asian-Pacific/Eurasian, and Middle Eastern papers, I could still compare papers from regions of the world that contribute the most to the English language computer science education literature—North America, Europe, and Asia-Pacifica/Eurasia et al.—at a fine level of detail. (There was only one paper from an African institution, and none from South American institutions, in the analysis of the planned contrasts.)

Interrater Reliability

Tables 10 through 20 present the number of cases (out of 53) that could be used to calculate an interrater reliability statistic, the ϰm, and its 95% confidence intervals. In short, the interrater reliabilities were good or fair (i.e., greater than .6) for most variables; however, they were lower than .60 on seven variables: Kinnunen's categories; type of paper, if not dealing with human participants; literature review present; setting adequately described; procedure adequately described; and results and discussion separate. Five out of seven variables with low reliabilities concern report elements.

Table 10. *Interrater Reliabilities for General Characteristics Variables*

General characteristics	n	Kappa	Lower CI 95%	Upper CI 95%
Kinnunen's categories	53	.40	.27	.55
Valentine's categories	53	.62	.48	.75
Human participants	53	.81	.66	.96
Anecdotal	34	.94	.82	1.00
Type of 'other'	17	.56	.27	.80

Table 11. *Interrater Reliabilities for Research Methods Variables*

Research method	n	Kappa	Lower CI 95%	Upper CI 95%
Experimental/quasi-experimental	17	.88	.65	1.00
Random assignment	10	.70	.40	1.00
Explanatory descriptive	17	.65	.29	1.00
Exploratory descriptive	17	.88	.65	1.00
Correlational	17	1.00		
Causal-comparative	17	.88	.65	1.00

Table 12. *Interrater Reliabilities for Experimental Design Variables*

Type of experimental design	n	Kappa	Lower CI 95%	Upper CI 95%
One-group posttest-only	10	1.00		
Posttest with controls	10	.80	.40	.10
Pretest/posttest with controls	10	.80	.40	.10
Group repeated measures	10	.80	.40	.10
Multiple factor	10	1.00		
Single case	10	1.00		

Table 13. *Interrater Reliabilities for Independent Variables*

Type of independent variable used	n	Kappa	Lower CI 95%	Upper CI 95%
Student instruction	10	1.00		
Teacher instruction	10	1.00		
Mentoring	10	1.00		
Speakers at school	10	1.00		
Field trips	10	1.00		
Computer science fair/contest	10	1.00		

Table 14. *Interrater Reliabilities for Type of Dependent Variable Measured*

Type of dependent variable measured	n	Kappa	Lower CI 95%	Upper CI 95%
Attitudes (student or teacher)	15	1.00		
Achievement in computer science	15	.60	.20	1.00
Attendance	15	.87	.60	1.00
Other	15	.72	.33	1.00
Computer use	15	.87	.60	1.00
Students' intention for future	15	1.00		
Teaching practices	15	.87	.60	1.00
Achievement in core (non-cs) courses	15	1.00		
Socialization	15	1.00		
Program implementation	15	1.00		
Costs and benefits	15	1.00		

Table 15. *Interrater Reliabilities for Grade Level and Undergraduate Year*

Grade level of participant	n	Kappa	Lower CI 95%	Upper CI 95%
Grade level	9	.39	.02	.75
Undergraduate year	2	1.00		

Table 16. *Interrater Reliabilities for Mediating or Moderating Variables*

Mediating or moderating variable	n	Kappa	Lower CI 95%	Upper CI 95%
Mediating/moderating factor examined	15	.71	.33	1.00
Gender	6	1.00		
Nationality	6	1.00		
Aptitude (in computer science)	6	1.00		
Race/ethnic origin	6	1.00		
Disability	6	1.00		
Socioeconomic status	6	1.00		
Other	6	1.00		

Table 17. *Interrater Reliabilities for Type of Effect Size Reported Variables*

Type of effect size reported	n	Kappa	Lower CI 95%	Upper CI 95%
Effect size reported	15	1.0		
Raw difference	14	1.0		
Variability reported with means	9	1.0		
Correlational effect size	14	1.0		
Standardized mean difference	14	1.0		
Odds ratio	14	1.0		
Odds	14	1.0		
Relative risk	14	1.0		

Table 18. *Interrater Reliabilities for Type of Measure Used Variables*

Type of measure used	n	Kappa	Lower CI 95%	Upper CI 95%
Questionnaires	15	.72	.33	1.00
Reliability or validity information	6	1.00		
Grades	15	.87	.60	1.00
Teacher- or researcher-made tests	15	.72	.33	1.00
Reliability or validity information	5	.60	-.19	1.00
Student work	15	.60	.20	1.00
Existing records	15	.87	.60	1.00
Log files	15	.72	.33	1.00
Standardized tests	15	.87	.60	1.00
Reliability or validity information	1	1.00		
Interviews	15	.87	.60	1.00
Direct observation	15	1.00		
Reliability or validity information[a]				
Learning diaries	15	1.00		
Focus groups	15	1.00		
Other	15	.87	.60	1.00

[a]No interrater reliability cases available.

Table 19. *Interrater Reliabilities or Type of Inferential Analyses Variables*

Type of inferential analysis used	n	Kappa	Lower CI 95%	Upper CI 95%
Inferential analyses used	15	1.00		
Parametric analysis	4	1.00		
Measure of centrality and dispersion reported	2	1.00		
Correlational analysis	4	1.00		
Sample size reported	1	1.00		
Correlation or covariance matrices	1	1.00		
reported	4	1.00		
Nonparametric analysis	1	1.00		
Raw data summarized	1	1.00		
Small sample analysis				
Entire data set reported[a]	4	1.00		
Multivariate analysis				
Cell means reported[a]				
Cell sample size reported[a]				
Pooled within variance or covariance matrix reported[a]				

[a]No interrater reliaibility cases available.

Table 20. *Interrater Reliabilities for Report Element Variables*

Report element	n	Kappa	Lower CI 95%	Upper CI 95%
Abstract present	15	.87	.60	1.00
Problem is introduced	15	.87	.60	1.00
Literature review present	15	.47	.07	.87
Research questions/hypotheses stated	15	.60	.20	1.00
Purpose/rationale	15	.06	-.33	.47
Participants adequately described	15	.72	.33	1.00
Setting adequately described	15	.47	.07	.87
Instrument adequately described	1	1.00		
Procedure adequately described	15	.47	.07	.87
Results and discussion separate	15	.47	.07	.87

Aggregated Results

In this subsection I present the aggregate findings. Note that in tables of groups of binomial variables, the column marginals do not sum to the total because one or more attributes could have applied. For example, an article could have used mixed-methods and could have been an experimental and explanatory descriptive type of article at the same time.

General Characteristics

Forum where article was published

Figure 4, which presents again the information in Table 9 collapsed across years, is a pie chart of the relative proportions of articles included in the sample, by forum. Note that *Bulletin* is the label for the June and December issues of *SIGCSE bulletin; CSE* is the label for the journal—*Computer Science Education; JCSE* is the label for the *Journal of Computer Science Education Online; SIGCSE* is label for the *Proceedings of the SIGCSE Technical Symposium,* which is published in the March Issue of *SIGCSE Bulletin; ITiCSE* is the label for the *Proceedings of the Innovation and Technology in Computer Science Education Conference,* which is published in the September issue of *SIGCSE Bulletin; Koli* is the label for the *Koli Calling: Finnish/Baltic Sea Conference on Computer Science Education; ACE* is the label for the *Proceedings of the Australasian Computing Education Conference;* and *ICER* is the label for the *International Computer Science Education Research Workshop.* The three forums that had published the most articles from 2000-2005 (*SIGCSE, ITiCSE,* and *Bulletin*) are all publications that are published by ACM in *SIGCSE Bulletin.*

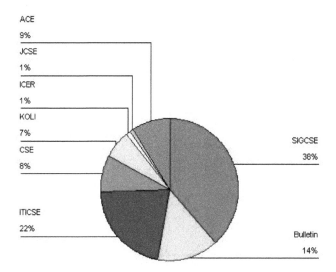

ACE
9%

JCSE
1%

ICER
1%

KOLI
7%

CSE
8%

ITICSE
22%

SIGCSE
38%

Bulletin
14%

Figure 4. A pie chart of percentage of articles in the sample by forum.

First authors whose articles were most frequently sampled

The first author whose articles were most frequently selected in this random sample was Ben-David Kollikant, with four articles. Other first authors whose articles were also frequently selected were A.T. Chamillard, Orit Hazzan, David Ginat, H. Chad Lane, and Richard Rasala, each with three articles in the sample.

First authors' affiliations

The authors of the articles in the selected sample represented 242 separate institutions. Of those 242 institutions, 207 were universities or colleges; 24 were technical universities, institutes of technology, or polytechnics; and 11 were other types of organizations, like research and evaluation institutes or centers. The majority of articles have first authors whom are affiliated with organizations in the U.S. or Canada. Table 21 shows the 12 institutions that were most often randomly selected into the sample. The number of articles that should correspond with the number of articles in the population can be estimated by multiplying the number of articles in the sample for each institution by 3.71, which is the ratio of the number of articles in the population to the

Table 21. *Institutions with Greatest Number of Articles*

Institution	Number of articles in sample	Proportion
University of Joensuu	13	3.7
Technion – Israel Institute of Technology	6	1.7
Drexel University	5	1.4
Northeasern University	5	1.4
Tel-Aviv University	5	1.4
Weizmann Institute of Science	5	1.4
Helsinki University of Technology	4	1.1
Michigan Technological University	4	1.1
Trinity College	4	1.1
University of Arizona	4	1.1
University of Technology, Sydney	4	1.1
Virginia Tech	4	1.1
Other institutions	289	82.4
Total	352	100.0

number of articles in the sample. The University of Joensuu, with 13 articles included in the sample, was an outlier. Of those 13 articles, 11 were from the Koli Conference, a conference held in a remote location near Joensuu.

Median number of authors per article

The median number of authors on each of the 352 articles was 2, with a minimum of 1 and a maximum of 7. The 2.5th and 95th percentiles of the median from 100,000 samples of size 352 were 5 and 5.

Median number of pages per article

Of the 349 articles that had page numbers, the median number of pages in the sample was 5, with a minimum of 3 and a maximum of 37. The 2.5th and 97.5th percentiles of the median from 10,000 samples of size 349 were 5 and 5.

Report elements

Table 22 shows the proportion of articles that had report elements that are considered by the American Psychological Association to be needed in empirical, behavioral papers.

Table 22. *Proportions of Report Elements*

Report element	n (of 123)	%	Lower CI 95%	Upper CI 95%
Abstract present	122	99.2	98.4	100.0
Problem is introduced	119	96.7	94.3	99.2
Literature review present	89	72.4	65.9	78.1
Purpose/rationale stated	45	36.6	30.8	42.3
Research questions/hypotheses stated	27	22.0	16.3	27.6
Participants adequately described	56	45.5	39.0	52.0
Setting adequately described	79	64.2	58.5	69.9
Instrument adequately described[a]	66	58.4	52.2	64.6
Procedure adequately described	46	37.4	30.9	43.9
Results and discussion separate	36	29.3	23.6	35.0

Note. Column marginals do not sum to 144 (or 100%) because more than one methodology type per article was possible. [a] Of 113.

Kinnunen's content categories

Table 23 shows how the articles were distributed according to Kinnunen's categories for describing the content of computer science education articles. Note that the interrater reliability for this variable was poor.

Valentine's research categories

Table 24 shows how the sampled articles were distributed into Valentine's research categories. Experimental and Marco Polo were the most frequently seen types of articles.

Human participants

Of the 352 articles in this sample, the majority of articles dealt with human participants. See Table 25.

Grade level of participants

Table 26 shows the grade level of participants of the 123 articles that dealt with human participants, that were not explanatory descriptive only, and that presented more than anecdotal evidence hereafter these 123 articles are called *the behavioral, quantitative, and empirical articles*). Bachelor's degree students were the type of participants most often investigated in the articles in this sample.

Table 23. *Proportions of Articles Falling into Each of Kinnunen's Categories*

Content category	n	%	Lower CI 95%	Upper CI 95%
New way to organize a course	175	49.7	45.7	54.0
Tool	66	18.8	15.3	22.2
Other	56	15.9	13.1	19.0
Teaching programming languages	31	8.8	6.5	11.4
Paraallel computing	10	2.8	1.4	4.3
Curriculum	5	1.7	0.6	2.8
Visualization	5	1.7	0.6	2.8
Simulation	2	0.6	0.0	1.1
Total	352	100.0		

Table 24. *Proportions of Articles Falling into Each of Valentine's Categories*

Valentine's category	n	%	Lower CI 95%	Upper CI 95%
Experimental	144	40.9	36.7	44.9
Marco Polo	118	33.5	29.7	37.5
Tools	44	12.5	9.7	15.3
Philosophy	39	11.1	8.5	13.6
Nifty	7	2.0	0.9	3.1
John Henry	0	0.0		
Total	352	100.0		

Table 25. *Proportion of Articles Dealing with Human Participants*

Human participants	n	%	Lower CI 95%	Upper CI 95%
Yes	233	66.2	62.2	70.1
No	119	33.8	29.8	37.8
Total	352	100.0		

Table 26. *Proportions of Grade Level of Participants*

Grade level of participant	n	%	Lower CI 95%	Upper CI 95%
Preschool	2	2.3	0.0	5.7
K-12	5	5.7	2.3	10.2
Bachelor's level	64	72.7	64.8	80.7
Master's level	1	1.1	0.0	3.4
Doctoral lavel	0	0.0		
Mixed level/other	16	18.2	11.4	25.0
Total	88	100.0		

As Table 27 shows, of the 64 Bachelor's degree participants, most were taking first-year computer science courses at the time the study was conducted. Studies in which the participants were not students (e.g., teachers) or the participants were of mixed grade levels were included in the mixed level/other category. (Note that the interrater reliability for the grade level of participants variable, but not the undergraduate year variable, was below a kappa of .4).

Anecdotal evidence only

Of the 233 articles that dealt with human participants, 38.2% presented only anecdotal evidence. See Table 28.

Types of articles that did not deal with human participants

Of the 119 articles that did not deal with human participants, the majority were purely descriptions of interventions. See Table 29, which shows the proportions of those articles that were program descriptions; theory, methodology, or philosophical papers; literature reviews; or technical papers. (Note that the interrater reliability estimate of kappa for this variable was below .6.)

Types of Research Methods and Research Designs Used

Types of research methods used

Table 30 shows that the experimental/quasi-experimental methodology type was the most frequently used type of methodology in the articles that dealt with human participants and that presented more than anecdotal evidence. Table 31 shows the proportions of quantitative articles

Table 27. *Proportion of Undergraduate Level of Computing Curriculum*

Year of undergraduate level computing curriculum	n	%	Lower CI 95%	Upper CI 95%
First year	39	70.9	61.8	80.0
Second year	3	5.5	1.8	90.9
Third year	8	14.5	7.3	2.2
Fourth year	5	9.1	3.6	14.6
Total	64	100.0		

Table 28. *Proportion of Human Participants Articles that Provide Anecdotal Evidence Only*

Anecdotal	n	%	Lower CI 95%	Upper CI 95%
Yes	89	38.2	33.1	43.3
No	144	61.8	56.7	66.5
Total	233	100.0		

Table 29. *Proportions of Types of Articles Not Dealing With Human Participants*

Type of article	n	%	Lower CI 95%	Upper CI 95%
Program description	72	60.5	53.8	67.2
Theory, methodology, or Philosophical paper	36	30.3	24.4	37.0
Literature review	10	8.4	5.0	11.8
Technical	1	0.8	0.0	1.7
Total	119	100.0		

Table 30. *Proportion of Methodology Types Used*

Methodology types	n	%	Lower CI 95%	Upper CI 95%
Experimental/quasi-experimental	93	64.6	58.3	70.8
Explanatory descriptive	38	26.4	20.8	31.3
Causal comparative	26	18.1	13.2	22.9
Correlational	15	10.4	7.0	14.6
Exploratory descriptive	11	7.6	4.2	11.1

Table 31. *Proportion of Types of Methods*

Type of method	n	%	Lower CI 95%	Upper CI 95%
Quantitative	107	74.3	68.1	80.2
Qualitative	22	15.3	10.4	20.8
Mixed	15	10.4	6.3	14.6
Total	144	100.0		

(i.e., not explanatory descriptive), qualitative articles (i.e., only explanatory descriptive), and mixed-methods articles (i.e., explanatory descriptive and one or more of the following: experimental/ quasi-experimental, exploratory descriptive, correlational, causal-comparative).

In terms of the 144 studies that dealt with human participants and that presented more than anecdotal evidence, convenience sampling of participants was used in 124 (86.1%) of the cases, purposive (nonrandom) sampling was used in 14 (9.7%) of the cases. Random sampling was used in 6 (4.2%) of the cases.

Research designs

Table 32 shows that the most frequently used research design was the one-group posttest-only design. Of the 51 articles that used that design, 46 articles used it exclusively (i.e., they did not use a one-group posttest-only design *and* a design that incorporated a pretest or a control of contrast group). In the sampled articles, quasi-experimental studies were much more frequently conducted than truly experimental studies. Of the 93 studies that used an experimental or quasi-experimental methodology, participants self-selected into conditions in 81 (87.1%) of the studies, participants were randomly assigned to conditions in 7 (7.5%) of the studies, and participants were assigned to conditions purposively, but not randomly, by the researcher(s) in 5 (5.4%) of the studies.

Independent, Dependent, and Moderating/ Mediating Variables Investigated

Independent variables

Table 33 shows the proportions of types of independent variables that were investigated in the 93 articles that used an experimental/quasi-experimental methodology. Nearly 99% of all independent variables were related to student instruction.

Table 32. *Proportions of Types of Experimental/Quasi-Experimental Designs Used*

Type of experimental design	n	%	Lower CI 95%	Upper CI 95%
Posttest only	51	54.8	47.3	62.4
posttest with controls	22	23.7	17.2	30.1
Pretest/posttest without controls	12	12.9	8.6	18.3
Repeated measures	7	7.5	4.3	11.8
Pretest/posttest with controls	6	6.5	2.2	10.8
Single-subject	3	3.2	1.1	5.3

Note. Column marginals do not sum to 93 (or 100%) because more than one methodology type per article was possible.

Table 33. *Proportion of Types of Independent Variables Used*

Type of independent variable used	n (93)	%	Lower CI 95%	Upper CI 95%
Teacher instruction	92	98.9	96.8	1.0
Mentoring	4	4.3	2.2	6.5
Speakers at school	2	2.2	0.0	5.3
Field trips	2	2.2	0.0	5.3
Computer science fair/contest	1	1.1	0.0	2.2

Note. Column marginals do not sum to 93 (or 100%) because more than one type of independent variable could have been used in each article (e.g., when there were multiple experiments).

Dependent variables

Table 34 shows the proportions of the different types of dependent variables that were measured in the 123 behavioral, quantitative, and empirical articles. Table 34 shows that attitudes and achievement in computer science were the dependent variables that were most frequently measured. The variables *project implementation* and *costs and benefits*, although included as categories on the coding sheet are not included in Table 34 because there were no studies that used them as dependent measures.

Mediating or moderating variables examined

Of the 123 behavioral, quantitative, and empirical articles; moderating or mediating variables were

Table 34. *Proportions of Types of Dependent Variables Measured*

Type of dependent variable measured	n (of 123)	%	Lower CI 95%	Upper CI 95%
Attitudes (student or teacher)	74	60.2	53.7	66.7
Achievement in computer science	69	56.1	49.6	62.6
Attendance	26	21.1	15.5	28.3
Other	14	11.5	7.4	15.6
Computer use	5	4.1	1.6	6.5
Students' intention for future	3	2.4	0.1	4.9
Teaching practices	2	1.6	0.0	3.3
Achievement in core (non-cs) courses	1	0.8	0.0	2.4
Socialization	1	0.8	0.0	2.4

Note. Column marginals do not sum to 123 (or 100%) because more than one type of dependent variables could have been measured.

Table 35. *Proportions of Mediating or Moderating Variables Investigated*

Mediating or moderating variable investigated	n (of 29)	%	Lower CI 95%	Upper CI 95%
Gender	6	20.7	13.8	27.6
Grade level[a]	4	13.8	6.9	20.7
Learning styles[a]	4	13.8	6.9	20.7
Aptitude (in computer science)[a]	2	6.8	3.5	10.3
Major/minor subject[a]	2	6.8	3.5	10.3
Race/ethnic origin	2	6.8	3.5	10.3
Age[a]	1	3.4	0.0	6.9
Amount of scaffolding provided[a]	1	3.4	0.0	6.9
Frequency of cheating[a]	1	3.4	0.0	6.9
Pretest effects[a]	1	3.4	0.0	6.9
Programming language[a]	1	3.4	0.0	6.9
Type of curriculum[a]	1	3.4	0.0	6.9
Type of institution[a]	1	3.4	0.0	6.9
Type of computing laboratory[a]	1	3.4	0.0	6.9
Type of grading (human or computer[a])	1	3.4	0.0	6.9
Self-efficacy[a]	1	3.4	0.0	6.9

Note. Column marginals do not sum to 29 (or 100%) because more than one methodology type per article was possible.
[a]These items were not a part of the original coding categories.

examined in 29 (23.6%). Table 35 shows the types and proportions of moderating or mediatingvariables that were examined in the sample of articles. There were many articles that examined moderating or mediating variables that fit into the *other* category (i.e., they were not originally on the coding sheet); those other variables were tabulated and have been incorporated into Table 35. Although included on the coding sheet, the variables—*disability* and *socioeconomic status*—were not included in Table 34 because no study examined them as mediating or moderating variables.

Types of Measures and Statistical Practices

Types of measures used

Table 36 shows the proportions of types of measures that were used in the 123 behavioral, quantitative, and empirical articles. Note that questionnaires were clearly the most frequently used type of measure. Measurement validity or reliability data were provided for questionnaires in 1 of 65 (1.5 %) of articles, for teacher- or researcher-made tests in 5 of 27 (18.5 %) of articles, for direct observation (e.g., interobserver reliability) in 1 of 4 (25%) of articles, and for standardized tests in 6 of 11 (54.5%) of articles.

Type of inferential analyses used

Of the 123 behavioral, quantitative, and empirical articles, inferential statistics were used in 44 (35.8%) of them. The other 79 articles reported quantitative results, but did not use inferential analyses. Table 37 shows the types of inferential statistics used, their proportions, and the proportion of articles that provided statistically adequate information along with the inferential statistics that were reported.

Type of effect size reported

Of the 123 behavioral, quantitative, and empirical articles, 120 (97.6%) reported some type of effect size. In the three articles that reported quantitative statistics but not an effect size, those articles presented only probability values or only reported if the result was "statistically significant" or not. Table 38 presents the types of effect sizes that were reported and their proportions. Odds, odds ratio, or relative risk were not reported in any of the articles in this sample. Of the articles that reported a raw difference effect size, 74 of those reported the raw difference as a difference

Table 36. *Proportions of Types of Measures Used*

Type of measure used	n (of 123)	%	Lower CI 95%	Upper CI 95%
Questionnaires	65	52.8	46.3	59.4
Grades	36	29.3	23.6	35.0
Teacher- or researcher-made tests	27	22.0	16.3	27.6
Student work	22	17.9	13.0	23.6
Existing records	20	16.3	11.4	21.1
Log files	15	12.2	8.1	9.2
Standardized tests	11	8.9	4.9	13.0
Interviews	8	6.5	3.3	9.8
Direct observation	4	3.3	0.8	5.7
Learning diaries	4	3.3	0.8	5.7
Focus groups	3	2.4	0.8	4.9

Note. Column marginals do not sum to 123 because more than one measure per article was possible.

Table 37. *Proportions of Types of Inferential Analyses Used*

Type of inferential analysis used	n	%	Lower CI 95%	Upper CI 95%
Parametric analysis (of 44)	25	56.8	47.7	65.9
Measure of centrality and dispersion				
Reported (of 25)	15	60.0	48.0	72.0
Correlational analysis (of 44)	13	29.5	23.3	37.2
Sample size reported (of 13)	10	76.9	53.9	92.3
Correlation or covariance matrix reported (of 13)	5	38.5	15.4	61.5
Nonparametric analysis (of 44)	11	25.0	13.2	31.8
Raw data summarized (of 11)	8	72.7	45.6	90.9
Small sample analysis (of 44)	2	4.5	0.0	9.1
Entire data set reported (of 2)	0	0.0		
Multivariate analysis (of 44)	1	2.3	0.0	2.3
Cell means reported (of 1)	0	0.0		
Cell sample size reported (of 1)	0	0.0		
Pooled within variance or covariance				
Matrix reported (of 1)	0	0.0		

Note. Column marginals do not sum because more than one methodology type per article was possible.

Table 38. *Proportions of Types of Effect Sizes Reported*

Type of effect size reported	n (of 123)	%	Lower CI 95%	Upper CI 95%
Raw difference	117	97.5	95.0	100.0
Correlational effect size	8	6.7	3.3	6.7
Standardized mean difference	6	5.0	1.7	8.3

Note. Column marginals do not sum to 120 (or 100%) because more than one methodology type per article was possible.

between means (the rest were reported as raw numbers, proportions, means, or medians). Of the 74 articles that reported means, 29 (62.5%) did not report a measure of dispersion along with the mean. Note that a liberal definition of a raw difference–also referred to as *relative risk* or a *gain score*—was used here. The authors did not actually have to subtract pretest and posttest raw scores (or pretest and posttest proportions) from one another to be considered a raw difference effect size. They simply had to report two raw scores in such a way that a reader could subtract one from another to get a raw difference.

Islands of Practice: Analysis of Crosstabulations

In this section I present the crosstabulated results for the 15 planned contrasts. Of the 15 contrasts, only the contrasts that were significant at the .003 probability level and the contrasts regarding the difference between articles published in papers and conferences are discussed in detail here. However, I do present crosstabulations for each of the 15 contrasts. Note that the probability level that corresponds with an overall probability level across the 15 contrasts of .05 is .003; see Stevens, 1999.

Differences between Journal and Conference Proceedings Articles

The results of these crosstabulation analyses show that there were no statistically significant differences between journal and conference proceedings articles in terms of several methodological attributes. Those attributes were the proportion of articles that provided anecdotal-only evidence, the proportion of articles that used an experimental or quasi-experimental method, the proportion of articles that used an explanatory descriptive method, the proportions of articles that used a one-group posttest-only research design exclusively, and the proportion of articles that examined attitudes as the only dependent variable. However, using the logistic regression approach it was found that there was a statistically significant difference, at the .10 alpha level, in the proportion of experimental/ quasi-experimental articles when a forum type by region interaction term in included in the model.

Anecdotal-only articles

Table 39 presents the frequencies and percentages of articles that dealt with human participants but only presented anecdotal evidence. The journal articles in this sample had 8.8% more anecdotal-only articles than conference articles; the difference in the overall observed cell deviations from the expected cell deviations was not statistically significant, $\chi^2(1, N = 233) = 1.32$, $p = .251$; resampled $p = .256$. In the case of Table 39, the adjusted residuals are small, which is congruent with the finding that χ^2 was not statistically significant. According to Agresti, "an adjusted residual that exceeds about 2 or 3 in absolute value indicates lack of fit (of the null hypothesis) in that cell" (1996, pp. 31-32).

Experimental/ quasi-experimental articles

Table 40 presents the frequencies and percentages of articles that reported on experimental or quasi-experimental investigations. Journal articles had 4.1% more experimental or quasi-experimental investigations than did conference articles; the difference between journal articles and conference articles was not statistically significant, $\chi^2(1, N=144) = 0.16$, $p = .687$; resampled $p = .672$. (See the logistic regression approach section for an alternate finding when a region by forum type interaction is controlled for.)

Explanatory descriptive articles

Journal articles had 7.1% more explanatory descriptive articles than did articles published in conference proceedings. This difference was not statistically significant, $\chi^2 (1, N=144) = 0.59$, $p = .441$; resampled $p = .426$. (See Table 41.)

Table 39. *Crosstabulation of Anecdotal-Only Papers in Conferences and Journals*

Forum	Anecdotal-only		Total	Percentage yes	Adjusted residual
	Yes	No			
Conference	66	116	182	36.3	-1.1
Journal	23	28	51	45.1	1.1
Total	89	144	233	38.2	

Table 40. *Crosstabulation of Experimental Papers in Conferences and Journals*

Forum	Experimental		Total	Percentage Yes	Adjusted residual
	Yes	No			
Conference	74	42	116	63.8	-0.4
Journal	19	9	28	67.9	0.4
Total	93	51	144	64.6	

Table 41. *Crosstabulation of Explanatory Descriptive Papers in Conferences and Journals*

Forum	Explanatory descriptive		Total	Percentage Yes	Adjusted residual
	Yes	No			
Conference	29	87	116	25.0	-0.8
Journal	9	19	28	32.1	0.8
Total	38	106	144	26.4	

Attitudes-only articles

Table 42 indicates that journals had 5.9% less articles that examined only attitudes than conference proceedings. The difference was not statistically significant, $\chi^2(3, N = 123) = 0.31$, $p = .580$; resampled $p = .579$.

One-group posttest-only articles

Table 43 shows the proportions of conference and journal articles that used one-group posttest-only research designs only and those that used designs with controls. Conference proceedings had 2.6% more articles that used the one-group posttest-only design exclusively than

Table 42. *Crosstabulation of Attitudes-Only Papers in Conferences and Journals*

Forum	Attitudes-only		Total	Percentage yes	Adjusted residual
	Yes	No			
Conference	32	68	100	32.0	0.6
Journal	6	17	23	26.1	-0.6
Total	38	85	123	30.9	

Table 43. *Crosstabulation of Experimental Papers That Used Posttest-Only Designs Exclusively*

Forum	Posttest-only exclusively		Total	Percentage yes	Adjusted residual
	Yes	No			
Conference	37	37	74	50.0	0.2
Journal	9	10	19	47.4	-0.2
Total	46	47	93	49.5	

did journal articles. The difference was not statistically significant, $\chi^2(1, N = 93) = 0.04, p = .838$; resampled $p = .835$.

Yearly Trends

Out of the five planned contrasts involving yearly trends, two were statistically significant. The number of anecdotal articles and the number of explanatory descriptive articles had decreased from 2000 to 2005.

Anecdotal-only articles

Table 44 shows that there was a decreasing trend in the number of anecdotal-only articles from 2000-2005. The fact that the adjusted residuals in the *Percentage Yes* column transition, more or less, from large positive values in 2000 to large negative values in 2005 and that the percentages, more or less, transition from larger to smaller support the finding that there was a trend. The trend was statistically significant, $M^2(1, N = 233) = 9.00, p = .003$; resampled $p = .003$.

66

Table 44. *Anecdotal-Only Papers by Year*

Year	Anecdotal-only		Total	Percentage yes	Adjusted residual
	Yes	No			
2000	18	13	31	58.1	2.4
2001	15	15	30	50.0	1.4
2002	9	17	26	34.6	-0.4
2003	14	25	39	35.9	-0.3
2004	18	34	52	34.6	-0.6
2005	15	40	55	27.3	-1.9
Total	89	144	233		

Explanatory descriptive articles

Table 45 shows that there was a somewhat decreasing trend in the number of explanatory descriptive articles that were published each year. Although the trend was not consistent (2002 was an exception to the trend), it was statistically significant, $M^2(1, N = 144) = 11.54$, $p = .001$; resampled $p < .000$.

Other types of articles

Crosstabulations for the types of articles where there was not a statistically significant trend (i.e., experimental/quasi-experimental articles, one-group posttest-only articles, and attitudes-only articles) are presented below. Table 46 shows that there was not a strong trend in the number of experimental/quasi-experimental papers that were published each year. Likewise for Table 47, which shows the number of one-group posttest-only articles per year, and for Table 48, which shows the number of attitudes-only papers by year. There was not strong evidence that there was a trend between the years 2000 and 2005.

Region of First Author's Affiliation

Of the five contrasts that dealt with the region of first author's affiliation, three were statistically significant. The statistically significant findings are described below.

Table 45. *Explanatory Descriptive Papers by Year*

Year	Explanatory descriptive		Total	Percentage yes	Adjusted residual
	Yes	No			
2000	7	6	13	53.8	2.4
2001	4	11	15	26.7	0.0
2002	8	9	17	47.1	2.1
2003	7	18	25	28.0	0.2
2004	9	25	34	26.5	0.0
2005	3	37	40	7.5	-3.2
Total	38	106	144		

Table 46. *Experimental/Quasi-Experimental Papers by Year*

Year	Experimental		Total	Percentage yes	Adjusted residual
	Yes	No			
2000	8	5	13	61.5	-0.2
2001	11	4	15	73.3	0.7
2002	10	7	17	58.8	-0.5
2003	14	11	25	56.0	-1.0
2004	22	12	34	64.7	0.0
2005	28	12	40	70.0	0.8
Total	93	51	144		

Note. $M^2(1, N = 144) = 0.17, p = .676$; resampled $p = .676$.

Table 47. *One-Group Posttest-Only Papers by Year*

Year	Anecdotal-only		Total	Percentage yes	Adjusted residual
	Yes	No			
2000	6	2	8	75.0	1.5
2001	6	5	11	54.5	0.4
2002	4	6	10	40.0	-0.6
2003	4	10	14	28.6	-1.7
2004	15	7	22	68.2	2.0
2005	11	17	28	39.3	-1.3
Total	46	47	93		

Table 48. *Attitudes-Only Papers by Year*

Year	Attitudes-only		Total	Percentage yes	Adjusted residual
	Yes	No			
2000	1	8	9	11.1	-1.3
2001	6	7	13	46.2	1.3
2002	3	9	12	25.0	-0.5
2003	5	17	22	22.7	-0.9
2004	12	17	29	41.4	1.4
2005	11	27	38	28.9	-0.3
Total	38	85	123		

Note. $M^2(1, N = 93) = 0.97, p = .326$; resampled $p = .315$.

Experimental/quasi-experimental articles

Table 49 shows that first authors who were affiliated with institutions in North America tend to write, and get published, articles that used experimental or quasi-experimental articles. In contrast, first authors who were affiliated with institutions in Europe or in the Middle East tended *not* to write, or get published, experimental or quasi-experimental articles. In fact, the odds of a first author affiliated with a North American association having published an experimental paper were more than 3.6 times greater than a first author affiliated with a European institution and more than 7.5 times greater than a first author affiliated with a Middle Eastern institution. The differences between observed and expected cell values in Table 49 were statistically significant, $\chi^2(3, N = 143) = 15.54$, $p = .001$; resampled $p < .000$.

Explanatory descriptive articles

Table 50 shows that first authors who were affiliated with a Middle Eastern institution tended to write and get published explanatory descriptive articles. The odds of a first author affiliated with a Middle Eastern institution having written and gotten published an explanatory descriptive articles was more than 13 times greater than the odds of their counterpart affiliated with a North American institution having written and gotten published an explanatory descriptive article. The differences were statistically significant, $\chi^2(3, N = 143) = 20.13, p < .000$; resampled $p < .000$.

Table 49. *Experimental Papers by Region of First Author's Affiliation*

Region	Experimental/ Quasi-experimental		Total	Percentage Yes	Adjusted residual
	Yes	No			
Eurasia	20	10	30	66.7	0.3
Europe	14	16	30	49.7	-2.3
Middle East	4	9	13	30.8	-2.6
North America	54	16	70	77.1	3.1
Total	92	51	143		

Table 50. *Explanatory Descriptive Papers by Region of First Author's Affiliation*

Region	Explanatory descriptive		Total	Percentage yes	Adjusted residual
	Yes	No			
Eurasia	5	25	30	16.7	-1.4
Europe	9	21	30	30.0	0.5
Middle East	10	3	13	76.9	4.3
North America	14	56	70	20.0	-1.7
Total	38	105	143		

Attitudes-only articles

Table 51 shows that the odds of a first author affiliated with an institution in the Asian Pacific or Eurasia having written and published an article in which attitudes were the sole dependent measure were more than 12 times greater than a first author affiliated with an institution in Europe. The differences were statistically significant, $\chi^2(3, N = 122) = 17.39, p = .00$; resampled $p < .000$.

Other types of articles

Crosstabulations for the types of articles in which there were no statistically significant regional differences (i.e., anecdotal-only papers and one-group posttest-only papers) are presented in Tables 52 and 53 below. (Note that the logistic regression analysis, however, showed that region is a statistically significant predictor of an article being an anecdotal-only article when the other factors are controlled for.)

Table 51. *Attitudes-only Papers by Region of First Author's Affiliation*

Region	Attitudes-only		Total	Percentage yes	Adjusted residual
	Yes	No			
Eurasia	16	10	26	61.5	3.9
Europe	3	24	27	11.1	-2.5
Middle East	1	4	5	20.0	-0.5
North America	17	47	64	26.9	-1.0
Total	37	85	122		

Table 52. *Anecdotal-Only Articles by Region of First Author's Affiliation*

Region	Anecdotal-only		Total	Percentage yes	Adjusted residual
	Yes	No			
Eurasia	10	30	40	25.0	-1.9
Europe	14	30	44	31.8	-1.0
Middle East	5	13	18	27.8	-.9
North America	59	70	129	45.7	2.7
Total	88	143	231		

Note. $\chi^2(3, N = 231) = 7.65$, $p = .054$; resampled p $= .059$.

Table 53. *One-Group Posttest-Only Papers by Region of First Author's Affiliation*

Region	One-group posttest-only		Total	Percentage yes	Adjusted residual
	Yes	No			
Eurasia	13	7	20	65.0	1.6
Europe	8	6	14	57.1	0.7
Middle East	3	1	4	75.0	1.1
North America	21	33	54	38.9	-2.3
Total	45	47	92		

Note. $\chi^2(3, N = 92) = 5.71$, $p = .127$; resampled p $= .128$.

For each of the five outcome variables (i.e., anecdotal-only papers, experimental/ quasi-experimental papers, explanatory descriptive papers, attitudes-only papers, and one-group posttest-only papers), I present the history of model fitting, information about the overall fit of the regression equation, and the regression equation(s) themselves. I also present graphs that visually portray the best fitting model. Note that the regression equations refer to probability of a *yes* (successful) outcome (i.e., p, not q).

On all of the outcomes besides explanatory descriptive, the African, Asia-Pacific/Eurasian, and Middle Eastern categories were combined into a combined region category called Asian-Pacific/Eurasian et al. I called it Asian-Pacific et al. because most of the observations came from the Asian-Pacific/Eurasian regions. The breakdown of articles into each region is given for each analysis below. Note that only articles that dealt with human participants are included in these regression analyses. A South American category was not included because there were no South American articles that dealt with human participants in the sample.

Anecdotal-Only Articles

Table 54 shows comparisons of the fit of several logistic regression models using anecdotal-only papers, a binary variable, as the outcome. In this case the best fitting model was Model 9: intercept + region + year + region * year. For the anecdotal-only papers variable, the Omnibus Test of Model Coefficients was statistically significant, $\chi^2(7, N = 233) = 20.74$, $p = .001$, and the Hosmer and Lemeshow test was not statistically significant, $\chi^2(7, N = 233) = 2.97, p = .888$, which indicate that the overall fit of the model was appropriate.

Figure 5 shows the scatterplot of expected and observed probabilities. It has one outlier at coordinate (0.5, 0.2), which corresponds with the three 2001 Asian-Pacific/Eurasian et al. anecdotal-only articles that dealt with human participants. A regression analysis was conducted with those three articles removed; I do not present those results of that analysis here because they were negligibly different from the results when the outlying data point was included.

Table 55 shows the results of regression analysis for the anecdotal-only papers. The breakdown of the *n*-size of the region categories was 129, 60, and 44 for North American, Asian-Pacific/Eurasian et al., and European articles, respectively. For the Asian-Pacific/Eurasian

Table 54. *The Fit of Several Logistic Regression Models for Anecdotal-Only Papers*

Model	Predictors	Deviance (df)	Models compared	Difference (df)	p
1	I+R+Y+F+R*Y+R*F+Y*F+R*Y*F	286.84(11)	-	-	-
2	I+R+Y+F+R*Y+R*F+Y*F	287.93(9)	1 & 2	1.09(2)	.58
3	I+R+Y+F+R*Y+R*F	288.32(8)	2 & 3	0.39(1)	.53
4	I+R+Y+F+R*Y+Y*F	288.01(7)	2 & 4	0.31(2)	.86
5	I+R+Y+F+R*F+Y*F	293.50(8)	2 & 5	5.57(1)	.02
6	I+R+Y+F+R*Y	288.45(6)	4 & 6	0.44(1)	.51
7	I+R+Y+F+F*Y	293.50(5)	4 & 7	0.00(2)	.99
8	I+R+Y+F	294.27(4)	6 & 8	5.79(2)	.06
9	I+R+Y+R*Y	289.17(5)	6 & 9	0.72(1)	.40

Note. I = intercept, R = region, Y = year, F = forum type.

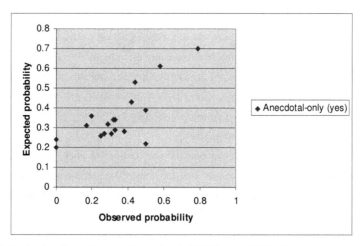

Figure 5. Scatterplot of expected and observed probability for anecdotal-only papers.

Table 55. *Summary of Regression Analysis for Predictors of Anecdotal-Only Articles, (N=233)*

Variable	B	S.E.	Wald	Df	p	Exp(B)
Year	-0.37	0.11	11.65	1	.00	.69
Region			9.65	2	.01	
North America (reference group)						
Asia-Pacific/Eurasia et al.	-2.24	0.79	7.95	1	.01	.11
Europe	-1.31	0.71	3.40	1	.07	.27
Region by year			5.33	2	.07	
North American (reference group)						
Asia-Pacific/Eurasia et al.						1.63
Europe	0.49	0.22	4.82	1	.03	1.30
	0.27	0.22	1.52	1	.22	2.33
Intercept						
	0.85	0.35	5.88	1	.02	

et al. category, the n-sizes for each region were 40, 18, and 2 for Asian-Pacific/Eurasian, Middle Eastern, and African articles, respectively.

The interpretation of logistic regression equations is as not as straightforward as it is for regression with a continuous outcome variable. Therefore, I will explain the interpretation of the items in the regression tables that are presented in this section. The first column shows the elements that were included in the regression equation; in the case of anecdotal-only papers those elements were a constant, year, region of first author's affiliation, and a region by year interaction. Because region was a categorical variable, the categories that it was comprised of—North America, Asia-Pacific/Eurasia et al., and Europe—are displayed. They are indented under the region label. In these regression analyses, North America was the reference group, so the comparisons were always be between North America and one of the other regions.

The second column, labeled B, shows the log coefficient. For a continuous variable, if the coefficient is positive, then that indicates that the odds of success (i.e., a yes) increase as the coefficient increases, and vice versa. For example, if the coefficient were positive for year, then that would indicate that the odds of a success would have increased every year. For categorical variables (like regions), the comparison category has a greater odds of success than the reference category if the log coefficient is positive, and vice versa. For example, if the coefficient for the Europe category were positive, that means that the likelihood of a European article's being an anecdotal-only article would have been greater than the likelihood of a North American article being an anecdotal-only article. If the coefficient were negative, the opposite would be true: The likelihood of a European article's being an anecdotal-only article would be less than the likelihood of a North American article's being an anecdotal-only article.

The column labeled S.E. displays the standard error of the log coefficient. The category labeled Wald shows the value of the Wald statistic, which, along with the degrees of freedom (df) in the next column, is used to determine the statistical significance of the coefficient.

Finally, since log coefficients alone cannot be easily interpreted, I have included the exponentiated B coefficient in the last column, labeled $exp(B)$. The value of 8 can be interpreted as an odds ratio—for categorical variables, the ratio of the odds in the reference category to the odds in the comparison category; for continuous variables, the ratio of odds between subsequent quantitative units. An odds ratio of one indicates that the odds of success are the same in both categories, an odds ratio less than one indicates that the odds are greater in the reference category, and an odds ratio greater than one indicates that the odds are greater in the comparison category. For example, an odds ratio of .27; where North America is the reference category, where Europe is the comparison category, and a success means that an article is anecdotal; would mean that the odds of a North American article's being anecdotal would be greater than for a European article—about 3.7 times greater because $1/.27 = 3.7$. If the odds ratios in the same case were 3.7 instead of .27, then that would mean that the odds in Europe papers were 3.7 times greater than the odds in North America papers.

So, based on the information given above, the following interpretations can be made from Table 55.

1. The predicted odds of an article's *not being anecdotal* had gotten 1.45 ($1/.69 = 1.45$) times greater per year between 2000 and 2005 (i.e., there was a decrease in anecdotal articles over time). The decrease was statistically significant.

2. The predicted odds of an article's being anecdotal were 9.1 ($1/.11 = 9.1$) times greater for North American articles than for Asian-Pacific/Eurasian et al. articles and 3.7 ($1/.27 = 3.7$) times greater for European articles. The difference between North America and Asian-Pacific/Eurasian et al. categories was statistically significant, and the difference between North American and European categories was nearly statistically significant ($p = .07$).

3. There was a statistically significant interaction in the difference between the decline in trend in anecdotal articles between North American articles and Asian-Pacific/Eurasian et al. articles.

Figure 6 shows the percentage of anecdotal-only articles to anecdotal-only plus nonanecdotal-only articles by region and year. The values next to each marker in a series show the number of anecdotal articles in that region each year. In Figure 6 it is clear that the percentage of North American anecdotal-only articles had decreased linearly between 2000 and 2005. Figure 6 also shows that the percentage of European anecdotal-only articles had dropped 30% between 2000 and 2001 and then leveled off. It also shows that there was considerable variability in the percentage of Asia-Pacific/Eurasian et al. articles across years.

Figure 7 shows the proportions of anecdotal-only articles by region. As shown in Table 55, there was a higher percentage of North American anecdotal-only articles than the percentage of European anecdotal-only articles, which was, in turn, higher than the percentage of Asian-Pacific/Eurasian et al. anecdotal-only articles.

ExperimentalQuasi-experimental Articles

Table 56 shows a history of model selection for the experimental/quasi-experimental variable. The best fitting model in this case, Model 9, was: intercept + region + forum type. However, I chose Model 7 over Model 9 in this case because after running the regression equation for Model 9, it turned out that Model 9 was exactly specified (i.e., there was perfect prediction if the continuous variable—year—was not included). Although Model 7 was a slightly more complicated model than Model 9, it had approximately the same deviance as Model 9. The differences between the values of the region, journal, and journal by region coefficients were negligible between models 7 and 9, so I only present the results of Model 9 here. Figure 8 shows a scatter plot of the expected and observed probabilities for experimental/quasi-experimental articles.

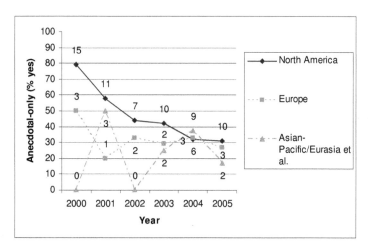

Figure 6. Line graph of anecdotal-only papers by combined region and year. The value nearest to a data point shows the n-size for that data point.

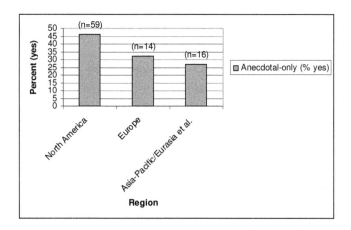

Figure 7. Bar graph of anecdotal-only papers by combined regions.

Table 56. *The Fit of Several Regression Models for Experimental/Quasi-Experimental Papers*

Model	Predictors	Deviance (df)	Models compared	Difference (df)	p
1	I+R+Y+F+R*Y+R*F+Y*F+R*Y*F	165.53(11)			
2	I+R+Y+F+R*Y+R*F+Y*F	167.10(9)	1 & 2	1.57(2)	.46
3	I+R+Y+F+R*Y+R*F	167.49(8)	2 & 3	0.39)1)	.53
4	I+R+Y+F+R*Y+Y*F	175.54(7)	2 & 4	8.44(2)	.01
5	I+R+Y+F+R*F+Y*F	168.93(7)	2 & 5	1.83(2)	.40
6	I+R+Y+F+R*Y	175.64(6)	3 & 6	8.15(2)	.02
7	I+R+Y+F+R*F	169.22(6)	3 & 7	1.73(2)	.42
8	I+R+Y+F	176.75(4)	7 & 8	7.53(2)	.02
9	I+R+F+R*F	169.31(5)	7 & 9	0.09(1)	.76

Note. I = intercept, R = region, Y = year, F = forum type.

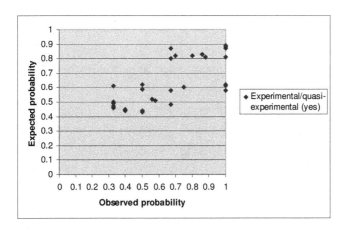

Figure 8. Scatterplot of expected and observed probabilities
for experimental/quasi experimental papers

The Omnibus Test of Model Coefficients was statistically significant, $\chi^2(6, N = 144)$ 17.89, p = .006, and the Hosmer and Lemeshow test was not statistically significant, $\chi^2(8, N = 144)$ 1.94, p = .983, which indicate that the overall fit of the model was good. There are three data points that I considered through visual analysis to be outliers, which are located approximately at coordinate (1.0, 0.6). Those data points represent the one nonanecdotal-only journal article from Europe in

2004, the three nonanecdotal-only journal articles from North America in 2004, and the one nonanecdotal-only journal article from North America in 2005. I ran regression equations with and without those outliers removed. The differences were minimal between the two equations so I only include the one with outliers here. The only notable difference however was that the *p*-value associated with forum type was .05 without outliers, and .09 with outliers (as shown in Table 57).

Table 57 shows a summary of the regression analyses when run with outliers. With outliers included, the breakdown of the *n*-size of the region categories was 70, 44, and 30 for North American, Asian-Pacific/Eurasian et al., and European articles, respectively. For the Asian-Pacific/Eurasian et al. category the breakdown of the *n*-sizes into regions was 30, 13, and 1 for Asian-Pacific/Eurasian, Middle Eastern, and African articles, respectively.

To illustrate the effect of the region by forum interaction, I also include the results of the regression equation without the region by forum interaction (with the outliers included) in 57. By comparing Tables 57 and 58 one can see that it is including the region by forum type interaction that causes the direction to switch on the forum type variable. Note that the model fit was statistically significantly better for the regression equation with the interaction term than without it (see Table 56). Yet, the regression equation without the interaction term had an overall good fit; the Omnibus Test of Model Coefficients was significant, $\chi^2(4, N = 144) = 10.49, p = .03$, and the Hosmer Lemeshow test was not significant, $\chi^2(8, N = 144) = 8.45, p = .390$.

The findings from these regression analyses, which are based on the regression equation with the outliers and interaction term left in, are listed below:

1. Region was a significant predictor of an article's being experimental/quasi-experimental or not. Specifically, the predicted odds of a North American article's being an experimental/quasi-experimental article were 4.6 (1/.22) times greater than an Asian-Pacific/Eurasian et al. article's odds and 5.6 (1/.18) times greater than the odds of European article's odds.

2. When controlling for the journal by region interaction, the odds of a conference article's being an experimental/quasi-experimental article were about 2.9 times (1/.34) greater than a journal article's odds.

3. There was a statistically significant interaction between type of forum and region.

Table 57. *Summary of Regression Analysis for Predictors Experimental/Quasi-Experimental Articles (N = 144), With Outliers*

Variable	B	S.E.	Wald	df	p	Exp(B)
Year	0.04	0.12	0.09	1	.77	1.04
Region			13.66	2	.00	
North America (reference group)						
Asia-Pacific/Eurasia et al.	-1.50	0.48	9.66	1	.00	0.22
Europe	-1.73	0.54	10.46	1	.00	0.18
Forum type						
Conference (reference group)						
Journal	-1.08	0.64	2.85	1	.09	0.34
Region by forum						
Journal by North American (reference group)			6.38	2	.04	21.21
Journal by Asia-Pacific/Eurasia et al.	3.10	1.29	5.64	1	.02	5.56
Journal by Europe	1.72	1.19	2.09	1	.15	4.00
Contrast	1.39	0.53	6.88	1	.01	

Table 58. *Summary of Regression Analysis for Predictors of Experimental/Quasi-Experimental Articles (N = 144), With Outliers and Without Interaction Term*

Variable	B	S.E.	Wald	df	p	Exp(B)
Year	0.30	0.11	0.08	1	.79	1.03
Region			9.56	2	.01	
North America (reference group)						
Asia-Pacific/Eurasia et al.	-0.94	0.42	5.02	1	.03	.39
Europe	-1.34	0.47	8.27	1	.00	.26
Forum type						
Conference (reference group)						
Journal	.14	0.47	0.08	1	.77	1.15
Constant	1.09	0.48	5.13	1	.02	2.97

Figure 9 shows the percent (yes) and number of experimental/quasi-experimental articles by forum type and region. It shows that there was a higher proportion of experimental/quasi-experimental articles in conferences than in journals in North American papers, but the opposite holds true for European and Asia-Pacific/Eurasia et al. papers. An explanation for this interaction and for the fact that forum type is significant here, but not in the crosstabulation of Table 40, is given in the discussion section. In Figure 10 it appears that the proportion of experimental/quasi-experimental papers did not change significantly across years.

Explanatory Descriptive Papers

For explanatory descriptive papers, I did not combine regional categories because the n-sizes of each category were large enough to get a sensible regression each equation. (I did not have to group Asian-Pacific/Eurasian, Middle Eastern, and African papers together.) I did however exclude the one African paper that was not ancecdotal-only from this analysis. Table 59 shows the history of model fitting for explanatory descriptive papers. Model 8 (intercept + region + year) turned out to be the best fitting model.

Figure 11 shows the expected and observed probabilities for explanatory descriptive papers. The Omnibus Test of Model Coefficients was statistically significant, $\chi^2(4, N = 143) = 27.22$, $p = .000$, and the Hosmer and Lemeshow test was not statistically significant, $\chi^2(8, N = 143) = 4.99$, $p = .768$, which indicate that the overall fit of the model was appropriate. Through visual inspection, I did not consider any of the data points to be outliers.

Table 60 shows the regression equation for explanatory descriptive papers. The breakdown of the n-sizes of the region categories here was 70, 30, 30, and 13 for North American, Asian-Pacific/Eurasian, European, and Middle Eastern articles, respectively. The one African nonanecdotal article was not included in this analysis. For the Asian- Pacific/Eurasian et al. category the n-sizes were 20, 4, and 1 for Asian-Pacific/Eurasian, Middle Eastern, and African articles, respectively.

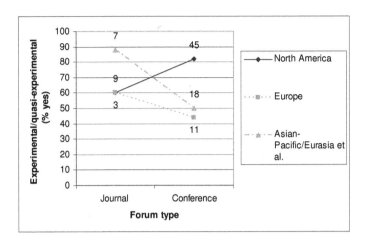

Figure 9. Line graph of experimental/quasi-experimental papers by combined region and forum type. The value nearest the data point shows the n-size for that data point.

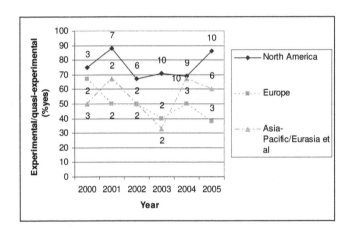

Figure 10. Line graph of experimental/quasi-experimental papers by combined region and year. The value nearest to a data point shows the n-size for that data point.

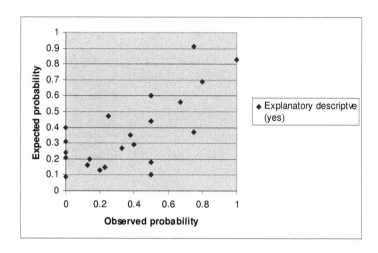

Figure 1. Scatterplot of Expected and Observed Probabilities for Explanatory Descriptive

Papers.

Table 59. *The Fit of Several Logistic Regression Models for Explanatory Descriptive Papers*

Model	Predictors	Deviance (df)	Models compared	Difference (df)	p
1	I+R+Y+F+R*Y+R*F+Y*F+R*Y*F	127.20(15)			
2	I+R+Y+F+R*Y+R*F+Y*F	130.79(12)	1 & 2	3.59(3)	.31
3	I+R+Y+F+R*Y+R*F	131.62(11)	2 & 3	0.83(1)	.36
4	I+R+Y+F+R*Y+Y*F	135.13(9)	2 & 4	4.34(3)	.23
5	I+R+Y+F+R*F+Y*F	132.49(9)	2 & 5	1.70(3)	.64
6	I+R+Y+F	138.30(5)	3 & 6	6.68(6)	.54
7	I+R+F	147.78(4)	6 & 7	9.48(1)	.00
8	I+R+Y	138.37(4)	6 & 8	0.07(1)	.79
9	I+Y+F	153.89(2)	6 & 9	15.59(2)	.00
10	I+R	147.78(3)	8 & 10	9.41(1)	.00
11	I+Y	153.96(1)	8 & 11	15.59(3)	.00

Note. I = intercept, R = region, Y = year, F = forum type.

Table 60. *Summary of Regression Analysis for Predictors of Explanatory Descriptive Articles, (N=143)*

Variable	B	S.E.	Wald	*p*	Exp(B)
Year	-0.39	0.13	8.91	.00	0.68
Region			13.00	.01	
North America (reference group)					
Asia-Pacific/Eurasia et al.	-0.17	0.59	0.08	.77	0.84
Europe	0.47	0.52	0.82	.36	1.60
Middle East	2.59	0.76	11.75	.00	13.31
Constant	-0.22	0.47	0.23	.63	0.80

The findings that relate to Table 60 are listed below:

1. Year was a significant predictor of explanatory descriptive papers. The odds of a paper's not being an explanatory descriptive paper was 1.47 (1/.68) times greater each year from 2000 to 20005.

2. Region was a significant predictor of a paper's being an explanatory descriptive paper. The odds of a Middle Eastern paper's being explanatory descriptive was over 13 times greater than the odds in a North American paper—a statistically significant difference in this case.

Figure 12 shows the percentage and number of explanatory descriptive papers by region. In Figure 12 there is considerable variability and low *n*-sizes. However, it appears that there had been a steady decrease in the number of North American explanatory descriptive papers from 2000 to 2005, although there was not a statistically significant interaction between year and region. Figure 13 shows the percentage and number of explanatory descriptive paper by region and year. The Middle Eastern category had the greatest proportion of explanatory descriptive papers.

Attitudes-Only Papers

Table 61 shows the history of model-fitting for attitudes-only papers. The best fitting model was actually Model 10 (intercept + region); however, I choose to keep *year* in the model because Model 10 was exactly specified. That is, I decided to use Model 8 (intercept + region + year) rather than Model 10. I ran logistic regressions for both Model 10 and for Model 8 and found that the

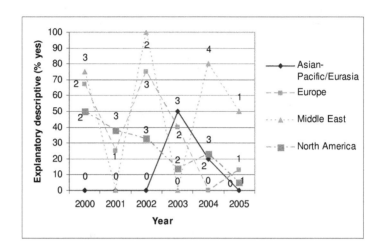

Figure 12. Line graph of explanatory descriptive papers by year and region. The value nearest to a data point shows the n-size for that data point.

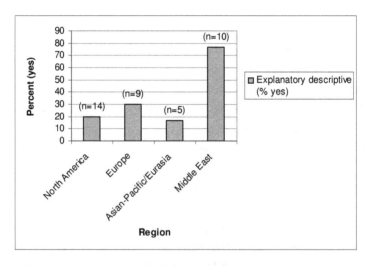

Figure 13. Bar graph of explanatory descriptive papers by region.

Table 61. *The Fit of Several Logistic Regression Models for Attitudes-Only Papers*

Model	Predictors	Deviance (df)	Models compared	Difference (df)	p
1	I+R+Y+F+R*Y+R*F+Y*F+R*Y*F	128.30(11)			
2	I+R+Y+F+R*Y+R*F+Y*F	129.33(9)	1 & 2	1.03(2)	.60
3	I+R+Y+F+R*Y+R*F	130.07(8)	2 & 3	0.74(1)	.39
4	I+R+Y+F+R*Y+Y*F	133.11(7)	2 & 4	3.78(2)	.15
5	I+R+Y+F+R*F+Y*F	132.93(7)	2 & 5	3.60(2)	.17
6	I+R+Y+F	136.05(4)	3 & 6	5.98(4)	.20
7	I+R+F	136.08(3)	6 & 7	0.03(1)	.86
8	I+R+Y	136.69(3)	6 & 8	0.61(1)	.44
9	I+F+Y	151.62(2)	6 & 9	15.57(2)	.00
10	I+R	136.79(2)	7 & 10	0.71(1)	.40
11	I+F	151.78(1)	7 & 11	15.70(2)	.00
12	I+Y	151.89(1)	8 & 12	15.20(2)	.00

Note. I = intercept, R = region, Y = year, F = forum type.

differences between them were negligible.

Figure 14 shows the expected and observed probabilities for attitudes-only papers. The Omnibus Test of Model Coefficients was statistically significant, $\chi^2(3, N = 123)$ 15.40, $p = .002$, and the Hosmer and Lemeshow test was not statistically significant, $\chi^2(8, N = 123) = 7.93$, $p = .440$, which indicates that the overall fit of the model was good.

Through visual inspection, I considered the data points at coordinates (0.7, 0.1) and (1.0, 0.55) to be outliers. The data point at coordinate (0.7,0.1) consisted of four articles from 2003 from the Asian-Pacific/Eurasian et al. category and the data point at coordinate (1.0, 055) consisted of three European articles from 2001. I ran regression analyses with and without the outliers and, because there was an interesting difference in the resulting regression equations, I present regression results for both.

Table 62 shows a summary of the regression analysis with outliers included and Table 63 shows a summary of the regression analysis with the outliers excluded. With outliers included, the breakdown of hte *n*-sizes of the combined region category was 64, 32, 27 for North American, Asian-Pracific/Eurasian et al., and European articles, respectively. For the Asian-Pacific/Eurasian et al. category, the *n*-sizes were 26, 5, and 1 for Asian-Pacific/Eurasian, Middle Eastern, and African articles, respectively.

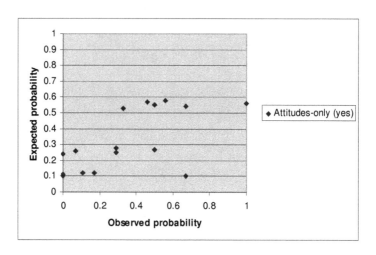

Figure 14. Scatterplot of expected and observed probabilities for attitudes-only papers.

Table 62. *Summary of Regression Analysis for Predictors of Attitudes-Only Articles (N = 123), With Outliers*

Variable	B	S.E.	Wald	df	p	Exp(B)
Year	.04	-0.13	0.10	1	0.75	1.04
Region			13.40	2	.00	
North America (reference group)						
Asia-Pacific/Eurasia et al.	1.27	0.46	7.77	1	.01	3.56
Europe	-1.06	0.68	2.44	1	.12	0.35
Constant	-1.16	0.54	4.71	1	.03	0.31

Table 63. *Summary of Regression Analysis for Predictors of Attitudes-Only Articles (N = 99), With Outliers Removed*

Variable	B	S.E.	Wald	Df	p	Exp(B)
Year	0.13	0.14	0.79	1	.37	1.14
			14.09	2	.00	
Region						
North America (reference group)						
Asia-Pacific/Eurasia et al.	1.28	0.46	7.81	1	.01	3.59
Europe	-2.13	1.06	4.04	1	.04	0.12
Constant	-1.45	0.57	6.40	1	.01	0.23

It was found that Region was a statistically significant predictor of an article's being an attitudes-only paper. The predicted odds of an Asian-Pacific/Eurasian article's being an attitudes-only article was 3.56 times higher than the predicted odds of a North American article's being an attitudes-only article. Also, the predicted odds of a European article's *not being* an attitudes-only articles was 2.9 (1/.35) times greater than predicted odds of a North American article's being an attitudes-only article.

Also, in the regression analysis with outliers excluded, the comparisons between the odds of both North American and Asian-Pacific/Eurasian et al. papers and between North American and European papers were statistically significant. In the regression analysis with the outliers included, the comparison of the odds between North American and Asian-Pacific/Eurasian et al. papers was statistically significant and the comparison between North American and European articles was nearly statistically significant (p = .12). Figure 15 shows the percentage of attitudes-only articles by year and combined region and Figure 16 shows the percentage of attitudes-only articles only by combined region. Those figures help illustrate the findings listed above: Namely, Asian-Pacific/Eurasian et al. articles had the higher proportion of attitudes-only articles.

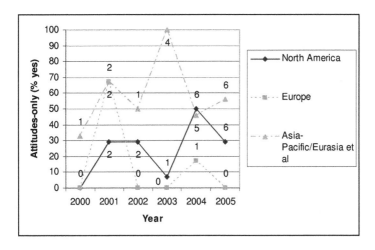

Figure 15. Line graph of attitudes-only papers by year and combined region. The value nearest to a data point shows the n-size for that data point.

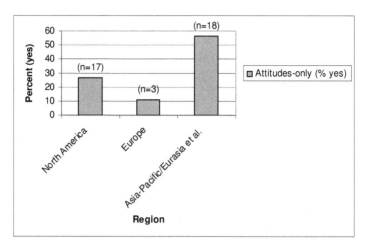

Figure 16. Bar graph of attitudes-only papers by combined regions.

One-Group Posttest-Only Articles

Table 64 shows the history of model-fitting for the one-group posttest-only articles. Based on Table 64, Model 9 (intercept + region + year + region by year) was the best model.

Figure 17 shows a plot of expected and observed probabilities (using Model 9) for one-group posttest-only articles. For Model 9, The Omnibus Test of Model Coefficients was statistically significant, $\chi^2(5, N = 93) = 14.53$, $p = .013$, and the Hosmer and Lemeshow test was not statistically significant, $\chi^2(8, N = 93) = 12.15$, $p = .15$, which indicate that the overall fit of the model was good.

I considered three data points to be outliers. They were approximately at coordinates (1.0, 0.65), (1.0, 5.5), (0.8, 3.5), and (0.55, .25); which correspond with the two experimental Asian-Pacific/Eurasian et al. articles in 2003, with the three experimental North American articles in 2001, with the nine experimental North American articles in 2003, and with the three experimental European articles in 2005. I ran regression analyses with and without outliers and found no meaningful differences whether outliers were included or not; therefore, I only present results here with the outliers included. Table 65 shows a summary of the regression analysis for Model 9. The breakdown of the *n*-size of the combined region category was 54, 25, 14 for North American, Asian-Pacific/Eurasian et al., and European articles, respectively. For the Asian-Pacific/Eurasian et al. category the *n*-sizes were 20, 4, and 1 for Asian-Pacific/Eurasian, Middle Eastern, and African articles, respectively.

Table 64. *The Fit of Several Logistic Regression Models for One-Group Posttest-Only Papers*

Model	Predictors	Deviance (df)	Models compared	Difference (df)	p
1	I+R+Y+F+R*Y+R*F+Y*F+R*Y*F	110.95(11)			
2	I+R+Y+F+R*Y+R*F+Y*F	113.00(9)	1 & 2	2.05(2)	.36
3	I+R+Y+F+R*Y+R*F	113.12(8)	2 & 3	0.12(1)	.73
4	I+R+Y+F+R*Y+Y*F	114.24(8)	2 & 4	1.24(1)	.27
5	I+R+Y+F+R*F+Y*F	120.48(7)	2 & 5	7.48(1)	.00
6	I+R+Y+F+R*Y	114.25(6)	3 & 6	1.13(1)	.29
7	I+R+Y+F+R*F	120.63(6)	3 & 7	7.51(1)	.00
8	I+R+Y+F	121.36(4)	6 & 8	7.11(2)	.03
9	I+R+Y+R*Y	114.39(5)	6 & 9	0.14(1)	.71
10	I+R+Y	121.79(3)	9 & 10	7.40(2)	.03

Note. I = intercept, R = region, Y = year, F = forum type.

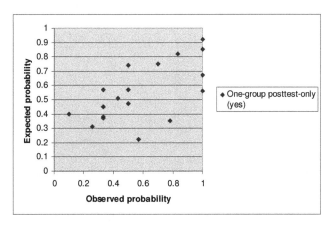

Figure 17. Scatterplot of expected and observed probabilities for one-group posttest-only articles with interaction term.

Table 65 shows that none of the predictor variables were significant predictors of one-group posttest-only papers. However, the interaction of year and region was statistically significant; specifically, there was an interaction between North American papers by year and Asian-Pacific/Eurasian papers by year. This interaction becomes clear from a visual examination of Figure 18, which is a graph of the percentages of one-group posttest-only papers by region and year.

In Figure 18, it shows that, more or less, there was a decline in the number of papers in Europe and North America. It also shows that, except for 2004, the pattern of decline of one-group

Table 65. *Summary of Regression Analysis for Predictors of One-Group Posttest-Only Articles for Model With Interaction Term (N= 93)*

Variable	B	S.E.	Wald	p	Exp(B)
Year	-0.21	0.18	1.44	.23	0.81
Region			2.99	.50	
North America					
(reference group; *n* = 54)					
Asia-Pacific/Eurasia et al. (*n* = 25)	-0.76	1.12	0.47	.50	0.47
Europe (*n* = 14)	2.23	1.66	1.97	.16	10.22
Region by year				.04	
North American (reference group)			6.38		
Asia-Pacific/Eurasia et al.	0.62	0.32	3.80	.05	1.86
Europe	-0.55	0.47	1.38	.24	0.58
Constant	0.24	0.63	0.14	.71	1.27

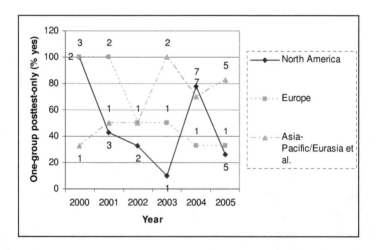

Figure 18. Line graph of one-group posttest-only articles by combined region. The value nearest to a data point shows the n-size for that data point.

posttest-only papers in Europe was similar to the pattern of decline in North America and that the North American series was usually slightly lower than in Europe. Also, Figure 18 shows that in the Asian Pacific et al. region there was an increase, except for 2004, in one-group posttest-only papers between 2000 and 2005. Although, Figure 18 indicates there was a difference between regions, the low n-sizes (only 5 out of 15 data points had n-sizes above 5) could have masked the difference in terms of finding statistical significance. Indeed, when collapsing across years, there was a statistically significant difference between regions, as Table 66 shows. Table 66, in which I show the results of Model 10—the regression equation without the interaction (i.e., intercept + region + year), shows that there was a statistically significant difference in the proportion of one-group posttest-only articles between North America and Asian-Pacific Eurasian et al. articles, but not between North American and European articles. This difference is also visualized in Figure 19, where the percentages of one-group posttest-only articles by region only are displayed. It is important to note, however, that Model 10 is not as good a fitting model as Model 9 (with the interaction) as Table 64 shows. Also, the Omnibus Test of Model Coefficients for Model 10, $\chi^2(3, N = 93) = 7.13$, $p = .068$, and the Hosmer and Lemeshow test, $\chi^2(7, N = 93) = 16.91$, $p = .018$, show that Model 9 is a poor model for predicting one-group posttest-only articles. Therefore, the results of Model 9 should be regarded with caution.

Comparisons between Fields

Up to this point I have presented results within the field of computer science education. In this section I present results concerning the proportions of empirical (i.e.,not anecdotal) articles dealing with human participants and proportions of quantitative, qualitative, and mixed methods research between fields. Note that the proportions for the field of education proper come from Gorard and Taylor (2004) and the proportions for the field of educational technology come from the review of methodological reviews of educational technology, which was presented earlier in this report.

Proportions of Empirical Articles Dealing with Human Participants

Table 67 shows that the proportions of empirical articles dealing with human participants decreased monotonically from education proper to educational technology and from educational technology to computer science education. Assuming that those fields are ordinal in terms of the degree to which they have an engineering tradition (where computer science education has the largest degree of the engineering tradition and education proper has the least), indicated by the number of articles that do not deal with human participants, the results of the M^2 test, indeed,

Table 66. *Summary of Regression Analysis for Predictors of One-Group Posttest-Only Articles for Model Without Interaction Term (N = 93)*

Variable	B	S.E.	Wald	p	Exp(B)
Year	-0.12	0.13	0.84	.36	.89
Region			5.85	.05	
North America					
(reference group; n = 54)					
Asia-Pacific/Eurasia et al. (N = 25)	1.21	0.51	5.55	.02	3.36
Europe (n = 14)	0.68	0.61	1.23	.27	1.98
Constant	-0.05	0.52	0.10	.92	0.95

Table 67. *Comparison of the Proportion of Empirical, Human Participants Articles in Computer Science Education and Education Proper*

Field	Empirical research with human participants		Total	Percentage yes	Adjusted residual
	Yes	No			
Ed. Proper	79	15	94	84.0	6.2
Ed. tech.	494	411	905	54.6	1.6
CSE	144	208	352	40.9	-5.3
Total	717	634	1,351		

Note. Ed. proper = education proper, Ed. tech. = educational technology, CSE = computer science education.

showed that there was a statistically significant linear (monotonic) relationship, $M^2(1, N = 1,351)$ = 52.32, p < .000. The adjusted residuals, which ranged from 6.2 for education proper and -5.3 for computer science education, showed that the linear relationship was pronounced.

Proportions of Types of Research Traditions between Fields

Table 68 shows that there was a statistically significant difference, $\chi^2(2, N = 638) = 20.84, p < .000$, between the proportions of quantitative, qualitative, and mixed methods articles in computer science education and educational technology forums. The adjusted residuals show that authors of computer science education articles tended to write, and get published, quantitative articles and

Table 68. *Comparison of the Proportion of Empirical, Human Participants Articles in Computer Science Education and Educational Technology*

Method	Field		Total	Percentage CSE	Percentage Ed. tech	Adjusted residual (CSE)
	CSE	Ed. tech.				
Quantitative	107	280	387	74.3	56.7	3.8
Qualitative	22	174	196	15.3	35.2	-4.6
Mixed	15	40	55	10.4	8.1	0.9
Total	144	494	638			

Note. CSE = computer science education, Ed. tech.= educational technology.

tended to not write, or get published, qualitative-only articles, compared to authors of papers published in educational technology forums. The percentage of mixed-method articles in each field was about the same however.

Table 69 shows that there was also a statistically significant difference, $\chi^2(2, N = 223) = 18.12$, $p < .000$, between the proportions of quantitative, qualitative, and mixed methods articles between the fields of computer science education and education research proper. The adjusted residuals show that the authors of computer science education research articles tended to use quantitative methods and tended to not use qualitative methods. Again, the proportions of mixed methods articles were about the same across fields.

Table 69. *Comparison of the Proportion of Empirical, Human Participants Articles in Computer Science Education and Education Proper*

Method	Field		Total	Percentage CSE	Percentage Ed. proper	Adjusted residual (CSE)
	CSE	Ed. proper				
Quantitative	107	43		74.3	54.4	3.0
Qualitative	22	32		15.3	40.5	-4.2
Mixed	15	4		10.4	5.1	1.4
Total	144	79				

Note. CSE = computer science education, Ed. proper. = education proper.

Chapter 4

Discussion of Findings

One limitation of this study was that the interrater reliabilities were low on a small proportion of the variables. I tried to circumvent this study limitation by not making strong conclusions about variables with poor reliabilities or by qualifying claims that were supported by variables with poor reliabilities. As was mentioned in the Methods section, I recognize that I approached this review from the viewpoint of a primarily quantitatively oriented behavioral science researcher. I investigated most deeply the quantitative experimental articles and did not deeply analyze articles that exclusively used explanatory descriptive modes of inquiry. Because of the significant variety and variability of explanatory descriptive methods, I was not confident that I could develop (or implement) a reliable system of classifying, analyzing, and evaluating those articles. Therefore, another study limitation was that I concentrated on experimental articles at the expense of explanatory descriptive articles. A third limitation had to do with the coders not being blind to certain characteristics of the articles (e.g., the institution, author, whether it came from a journal or a conference proceeding). Therefore, coder bias was possible. However, I have reasons to believe that coder bias did not unduly affect the results. The first is that because there was an interrater reliability coder, the coder bias would have had to have operated in the same direction for both coders, otherwise the interrater reliabilities would have been low. Although it is possible that both the primary and secondary coders had the same bias, it is less probable than just a single coder having the bias. Also, had there been coder bias, as I discuss in the section on the difference between journal and conference papers, the bias probably would have manifested itself in a way that supported the hypothesis. However, on the variables where coder bias would have been harmful, such as the difference between journals and conference proceedings, the results contradicted the hypothesis.

Interpretation of Descriptive Findings

My primary research question, which I addressed in terms of nine subquestions, was—What are the methodological properties of research reported in articles in major computer science education research forums from the years 2000-2005. A summary list of answers to each of those research questions is given below:

1. About one third of articles did not report research on human participants.
2. Most of the articles that did not deal with human participants were program descriptions.
3. Nearly 40% of articles dealing with human participants only provided anecdotal evidence.
4. Of the articles that provided more than anecdotal evidence, most articles used experimental/quasi-experimental or explanatory descriptive methods.
5. Questionnaires were clearly the most frequently used type of measurement instrument. Almost all of the measurement instruments that should have psychometric information provided about them did not have psychometric information provided.
6. Student instruction, attitudes, and gender were the most frequent independent, dependent, and mediating/moderating variables, respectively.
7. Of the articles that used an experimental research design, the majority used the one-group posttest-only design.
8. When inferential statistics were used, the amount of statistical information used was inadequate in many cases.

Because of the poor interrater reliabilities, I am hesitant about making summary conclusions about the types of articles that did not deal with human participants (related to Question 2) and about the question related to article structures (Question 9). In terms of my secondary research questions about islands of practice, I conducted 15 planned contrasts. Those 15 contrasts concerned the differences between journals and conference papers, yearly trends, and the regions of affiliation of the first authors, on the major methodological variables: proportion of anecdotal only papers, proportion of experimental/quasi-experimental papers, proportion of explanatory descriptive papers, proportion of papers using a one-group posttest-only design, and proportion of papers measuring attitudes only. The major findings about the islands of practice and trends in computer science education research are listed below:

9. There was no difference in major methodological characteristics between articles

published in computer science education journals and those published in peer-reviewed conference proceedings. However, there is some evidence that there was a slightly higher proportion of experimental/quasi-experimental articles in conference proceedings when a region by forum type reaction is controlled for.

10. There was a decreasing yearly trend in the number of anecdotal-only articles and in the number of articles that used explanatory descriptive methods.

11. First authors affiliated with North American institutions tended to publish papers in which experimental/quasi-experimental methods were used; first authors affiliated with Middle Eastern or European institutions tended to not publish papers in which experimental or quasi-experimental methods were used.

12. First authors affiliated with Middle Eastern institutions strongly tended to publish explanatory descriptive articles.

13. First authors affiliated with Asian-Pacific or Eurasian institutions tended to publish articles in which attitudes were the sole dependent variable; and

14. First authors affiliated with North American institutions tended to publish more anecdotal-only articles than their peers in other regions. However, this proportion had been decreasing linearly over time.

Proportion of Human Participants Articles

My prediction for the proportion of articles that would not report research on human participants; which was based on the Randolph, Bednarik, and Myller (2005); was between 80% and 60%. However, the proportion in the current review (33.8%) was about 30% lower than I had predicted. My explanation for this discrepancy is that the Koli forum, on which my prediction was based, simply had a higher proportion of research that did not deal with human participants than the computer science education research in general.

Proportion of Program Description Articles

Earlier I made a prediction that the majority of articles that would not deal with human participants would be program descriptions; that prediction was confirmed. Of the 34% of papers that did not report research on human participants, most (60%) of the papers were purely descriptions of interventions without any analysis of the effects of the intervention on computer science students. This proportion of articles is slightly higher, but near, the proportion of program descriptions in other computing-related methodological reviews in which the proportion of program descriptions was measured.

Assuming that Valentine's (2004) categories—Marco Polo and Tools—coincide with my program description category, then Valentine's findings are similar to my own; he found that 49% of computer science education research articles are what he called Marco Polo or Tools articles. In addition, Tichy and colleagues (1995) found that 43% of the computer science articles in their study were design and modeling articles, which would be called program descriptions in my categorization system. One of the assumptions of this report is that the proportion of program description-type articles is an indicator that the engineering tradition of computer science (see Tedre, 2006) is an artifact in computer science education research. Although it would be foolish to recommend an ideal proportion of program description and formalist articles to empirical articles dealing with human participants, perhaps a statement by Ely, one of the key figures in educational technology, can help inform the practice of computer science education. In an article in which Ely re-examined some of his assertions about the philosophy of educational technology made 30 years prior, he had the following to say about his earlier assertion that "the behavioral science concept of instructional technology is more valid than the physical science concept" (1999, p. 307):

> The original intent of this statement [that the behavioral science concept of instructional technology is more valid than the physical concept] was to contrast the psychology of learning (behavioral science) with the hardware/software aspects of technology (physical science). Using the same construct today, behavioral science becomes psychology of learning and instruction while physical science remains as the hardware/software configurations that deliver education and training. The psychological concept here is often referred to as instructional design (or sometimes, instructional systems design). There is growing evidence that the use of instructional design procedures and processes lead to improved learning without regard to the hardware and software that is used. Design is a more powerful influence on learning than the system that delivers it. (p. 307)

The conclusion I drew from this quote, which can also be applied to computer science education, is that while many computer science educators may be experts at creating the software and hardware to create automated interventions to increase the learning of computer science, an increased emphasis should be put on the instructional design of the intervention rather than only or primarily on the software and hardware mechanisms for delivering the instructional intervention merits careful consideration.

One way to inform the dialogue about the distributions of research methods in computer science education is to examine statements from authorities such as Ely or the variety of working groups on computer science education. Another way to inform the dialogue is to relate the research areas in computer science education to the types of research methods that are used in it.

In terms of the types of research areas in computer science education, there are several taxonomy systems that have been used. These include taxonomies presented in Fincher and Petre (2004), Glass and colleagues (2004), and Valentine (2004). Pears, Seidman, Eney, Kinnunen, and Malmi (2005) critically reviewed those taxonomies and concluded that Fincher and Petre's taxonomy of research areas was superlative because it "corresponded best to the diversity of computing education research" (p.154). Fincher and Petre's 10 research areas (as cited in Pears et al.) are listed below:

1. Student understanding.
2. Animation, visualization, and simulation.
3. Teaching methods.
4. Assessment.
5. Educational technology.
6. Transferring professional practice to the classroom.
7. Incorporating new developments and new technologies.
8. Transferring from campus-based teaching to distance education.
9. Recruitment and retention.
10. Construction of the discipline. (p. 153)

In terms of the types of research methods that are used in fields related to information technology, Jarvinen (2000) has proposed a useful taxonomy. In that taxonomy of research approaches, Jarvinen first divided the variety of research approaches into two classes: (a) approaches studying reality and (b) mathematical approaches. Jarvinen further divided the "approaches studying reality" category into five subcategories: (a) conceptual-analytical approaches, (b) theory-testing approaches, (c) theory-creating approaches, (d) artifacts-building approaches, and (e) artifacts-evaluating approaches. Now, relating Jarvinen's (2000) taxonomy of research approaches to Fincher and Petre's (2005) taxonomy of research areas, the relation between the distribution of research approaches and the major research areas becomes clearer. From my perspective, categories 1, 2, 3, 4, 6, 7, 8, 9, and the research component of Category 5—educational technology–lend themselves to empirical research with human participants. The development component of the educational technology category, in as much as that means the development of learning technologies, lends itself to what Jarvinen calls artifacts building approaches. I do not consider Fincher and Petre's "incorporating new developments and new technologies" research area to be an area that refers to the construction of new developments and technologies. I argue, rather, that it refers to the implementation of technologies into the physical learning environment, which is a research area that lends itself to empirical approaches that deal with human participants. If the majority of research areas in Fincher and Petre's (2005) taxonomy do lend themselves to empirical research approaches that deal with human participants, then it would make sense to assume that the majority of research approaches would be empirical research approaches that deals with human participants. Indeed, that was what was found in this methodological review: Over 66% of the research papers in this review used approaches that dealt with human participants (see Table 25). One interesting finding though was that there was such a large proportion of reports on artifact-building (i.e., what I called program descriptions) given that the artifacts-building approach was directly relevant in only 1 subcategory in 1 out of 10 of Fincher and Petre's categories—the development component of the educational technology category. In fact, about 21% (78/ 352) of the total articles sampled in this methodological review were purely program descriptions. The conclusion that I drew from this finding was that the research areas in Fincher and Petre's taxonomy are not equally represented in the computer science education research literature—it seems that the development component of the educational technology research area makes up a larger part of the computer science education literature than the other research areas.

In fact, the development component in the computer science education research literature makes up an even larger proportion than the developmental component in the educational technology research literature itself. Supposing that across the fields of educational technology and computer science education research there are equal proportions of program/tool descriptions in the articles that do not deal with human participants, then the proportion of program/tool descriptions in the computer science education research literature is almost 15% higher than in the field of education technology (see Table 69). This finding is surprising because one would assume that computer science education is a field characterized as largely technology education, not educational technology.

Proportions of Anecdotal-only Articles

The issue of the proliferation of anecdotal evidence in computing research, especially in software engineering, was being addressed over ten years ago. Holloway (1995) wrote:

> Rarely, if ever, are [empirical claims about software engineering] augmented with anything remotely resembling either logical or experimental evidence. Thus, one can conclude that software engineering is based on a combination of anecdotal experience and human authority. That is, we know that a particular technique is good because John Doe, who is an

authority in the field says that it is good (human authority); John Doe knows that it is good because it worked for him (anecdotal experience). Resting an entire discipline on such a shaky epistemological foundation is absurd, but ubiquitous nonetheless. (p. 21)

As Table 28 showed, the proliferation of anecdotal evidence is also an issue for the current computer science education research. The proportion of anecdotal-only articles was 22.3% higher than I had predicted based on previous research. Note that by the term anecdotal evidence in this review I have meant the informal observation of a phenomenon by a researcher. I do not necessarily mean that humans cannot make valid and reliable observations themselves, as happens in ethnographic research or research in which humans operationalize and empirically observe behavior.

Also, I concur that anecdotal experience has a role in the research process–it has a role in hypothesis generation. But, as Holloway (1995) pointed out, there are major problems to using informal anecdotal experience as the sole means of hypothesis confirmation. Valentine in his methodological review came to the same conclusion about the proliferation of anecdotal evidence in the field computer science education research. In fact, he ended his article with a call for more research not based on anecdotal experience. Valentine (2004) wrote:

We need more [conclusions that are based on defensible research, and not mere assumptions] of this in SIGCSE. I challenge the creators of CS1/CS2 Tools, in particular to step up and prove to us that your Tool actually does what you are claiming that it does. Do the fundamental research necessary to rest your claims upon defensible fact. (p. 259)

This sentiment about the importance of collecting empirical data is also echoed in several papers on computer science education research such as Clancy, Stasko, Guzdzial, Fincher, and Dale (2001) and Holmboe, McIver, and George (2001). Also concerning anecdotal evidence, it is important that computer science education researchers make claims that are congruent with the quantity and quality of evidence that was collected. For example, if a CSE researcher were to write "Our intervention caused students to learn more, more quickly" and the evidence that was collected consisted only of informal, anecdotal observations, then that would surely be an example of a mismatch between what was claimed and what, in the spirit of scientific honesty, should have been claimed. I did not code for a mismatch between a claim and what could have been claimed based on anecdotal evidence. However, based on my own anecdotal experience from reviewing about one quarter of the mainstream computer science education research published between 2000 and 2005, I hypothesize that this mismatch between claim and evidence for the claim does exist and that it is even common.

Types of Research Methods Used

I predicted that most articles that provided more than anecdotal evidence for their claims would use experimental/quasi-experimental or exploratory descriptive methods more than other methods. I was correct in the prediction that experimental/quasi-experimental methods would be used more frequently than other methods. However, I was wrong on the other part of the prediction; explanatory descriptive methods were used more often than exploratory descriptive methods. Perhaps this a good sign for the state of computer science education research; it signals a shift from the description of phenomena to the causal explanation of phenomena.

Experimental/quasi-experimental and explanatory descriptive methods are both methods that allow researchers to make causal inferences, and thereby confirm their causal hypotheses (Mohr, 1999). Experimental/quasi-experimental research is predicated on a comparison between a counterfactual and factual condition, via, what Mohr called, factual causal reasoning. Explanatory descriptive research is predicated on what Mohr called physical causal reasoning, or what Scriven (1976) called the Modus Operandi Method of demonstrating causality.

To illustrate the difference between these approaches, suppose that it is a researcher's task to prove that turning on the light switch in a room causes that room's light to come on. Using factual causal reasoning the researcher would conduct an experiment in which the researcher would note that when the switch is put in the "off" position, the light goes off (the factual condition); that when the switch is put in the "on" position, the light goes on (the counterfactual condition); and that the light never goes on unless the switch is in the on position, and vice versa—disregarding the possibility of a burnt-out bulb. Through this factual causal reasoning process of comparing factual and counterfactual conditions the researcher would arrive at the conclusion that turning the switch on causes the light to go on.

On the other hand, if the researcher were to use physical causal reasoning to determine if turning the switch on causes the light to come on, the process would be entirely different. The research might tear through the walls and examine the switch, the light, the power source, and the electrical wiring between the switch, the light, and the power source. By knowing the theory of how electricity and circuits work, the researcher, without ever having turned on the switch would be able to say with confidence that turning on the switch will cause the light to come on.

At any rate, the fact that most of the research being done in computer science education is done with types of methods that could possibly arrive at causal conclusions (given that the research is conducted properly) is a positive sign for computer science education research. Explanatory descriptive researchers in computer science education use physical causal reasoning to arrive at their causal conclusions; experimental researchers compare factual and counterfactual conditions. This fact indicates that computer science education researchers are asking causal questions and also choosing methods that can answer causal questions, if the method is conducted properly.

Types of Measures Used

Based on previous research I predicted that questionnaires, grades, and log files would be the most frequently used types of measures. I was correct except that teacher- or researcher-made tests were used more often than log files. Another prediction was that few or none of the measures that should have had psychometric information reported, had that information reported. This was especially true of questionnaires; only 1 out of 65 articles in which questionnaires were used gave any information about the reliability or validity of the instrument. According to Wilkinson et al., "if a questionnaire is used to collect data, summarize the psychometric properties of its scores with specific regard to the way the instrument is used in a population. Psychometric properties include measures of validity, reliability and internal validity" (1999, n.p). Obviously, the lack of psychometric information about instruments is a clear weakness in the body of the computer science education research.

Proportions of Dependent, Independent, and Mediating/Moderating Variables Examined

My prediction was that student instruction, attitudes, and type of course would be the most frequently used types of independent, dependent, and mediating/moderating variables, respectively. My prediction was correct. Mark Guzdial, one of the members of the working group on Challenges to Computer Science Education Research, admits that, "We know that student opinions are unreliable measures of learning or teaching quality" (Almstrum et al., 2005, p. 191). Yet, this review shows that attitudes are the most frequently measured variable. In fact, 44% of articles used attitudes as the sole independent article. While attitudes may be of interest to computer science education researchers, as Guzdial suggests, they are unreliable indicators of learning or teaching quality.

Experimental Research Design Used

I was correct in my prediction that the one-group posttest-only and posttest-only with control designs would be the most frequently used type of research designs. It is important to note that the one-group posttest-only design was used more than twice as often as the next most frequently used design, the posttest-only design with controls. Although the one-group posttest-only design is the most common experimental design in computer science education research, it is also probably the worst of the experimental research designs in terms of internal validity. According to Shadish et al. (2002), "nearly all threats to internal validity except ambiguity about temporal precedence usually apply to this design. For example a history threat is nearly always present because of other events might have occurred at the same time as the treatment" (p. 107). They do argue, however, that the [one-group posttest-only] design has merit in rare cases in which "much specific background knowledge exists about how the dependent variable behaves. . . For valid descriptive causal inferences to result, the effects must be large enough to stand out clearly, and either the possible alternative causes must be known and be clearly implausible or there should be no known alternative that could operate in the study context" (Campbell, 1975). These conditions are rarely met in the social sciences, and so this design is rarely useful in this simple form. (p. 107) The obvious conclusion is that the one-group posttest-only design is poor for making causal inferences in most cases. Other designs, with pretests and/or control groups, obviously would be better design choices if the goal is causal inference.

In terms of random selection and random assignment, I correctly predicted that these would be rare in the computer science education research. Convenience samples were used in 86% of articles, and students self-selected into treatment and control conditions in 87% of the articles. While some, such as Kish (1987) and Lavori, Louis, Bailar, and Polansky (1986), are staunch advocates of the formal model of sampling (i.e., random sampling followed by random assignment), there are others that question that model's utility. Shadish and colleagues (2002) claim that formal sampling methods have limited utility for the following reasons:

1. The [formal] model is rarely relevant to making generalizations to treatments and effects.
2. The formal model assumes that sampling occurs from a meaningful population, though ethical, political, and logical constraints often limit random selection to less meaningful populations.
3. The formal model assumes that random selection and its goals do not conflict with random assignment and its goals.
4. Budget realities rarely limit the selection of units to a small and geographically circumscribed population at a narrowly prescribed set of places and times.
5. The formal model is relevant only to generalizing to populations specified in the original sampling plan and not to extrapolating to populations other than those specified.
6. Random sampling makes no clear contribution to construct validity. . . (p. 348)

Shadish and colleagues (2002) concluded that "although we unambiguously advocate [formal random sampling] when it is feasible, we cannot rely on it as an all-purpose theory of generalized theory of causal inference. So researchers must use other theories and tools to explore generalized causal inference of this type" (p. 348). Some of the 'other theories and tools to explore generalized causal inference" are listed below:

1. Assessing surface similarity—"assessing the apparent similarities between study operations and the prototypical characteristics of the target population" (p. 357).

2. Ruling out irrelevancies—"identifying those attributes of persons, settings, treatments, and outcome measures that are irrelevant because they do not change a generalization" (p. 357).

3. Making discriminations—"identifying those features of persons, settings, treatments, or outcomes that limit generalization" (p. 357).

4. Interpolating and extrapolating—"generalizing by interpolating to unsampled values within the range of sampled persons, settings, treatments, and outcomes by extrapolating beyond the sampled range (p. 366).

5. Making causal explanation—developing and testing explanatory theories about the target of generalization (p. 366).

This same notion was expressed by Wilkinson et al. (1999). They stated:

> Using a convenience sample does not automatically disqualify a study from publication, but it harms your objectivity to try to conceal this by implying that you used a random sample. Sometimes the case for the representativeness of a convenience sample can be strengthened by explicit comparison of sample characteristics with those of a defined population across a wide range of variables. (n.p.)

The conclusion for computer science education researchers is that while random sampling is desirable when it can be done, doing purposive sampling or at least assessing the representativeness of a sample by examining surface similarities, ruling out irrelevancies, making discriminations, and interpolating and extrapolating, and examining causal explanations can be a reasonable alternative.

In terms of random assignment of participants to treatment conditions, the same types of lessons apply. While random assignment is desirable, when it is not feasible there are other ways to make strong causal conclusions. This is explained in Wilkinson et al. (1999):

For research involving causal inferences, the assignment of units to levels of the causal variable is critical. Random assignment (not to be confused with random selection) allows for the strongest possible causal inferences free of extraneous assumptions. If random assignment is planned, provide enough information to show that the process for making the actual assignments is random.

For some research questions, random assignment is not feasible. In such cases, we need to minimize effects of variables that affect the observed relationship between a causal variable and an outcome. Such variables are commonly called confounds or covariates. The researcher needs to attempt to determine the relevant covariates, measure them adequately, and adjust for their effects either by design or by analysis. If the effects of covariates are adjusted by analysis, the strong assumptions that are made must be explicitly stated and, to the extent possible, tested and justified. Describe methods used to attenuate sources of bias, including plans for minimizing dropouts, noncompliance, and missing data. (n.p.)

The conclusion for computer science education researchers is that when it is not possible to randomly assign participants to experimental conditions, steps need to be made, through design or analysis, to "minimize the effects of variables that affect the observed relations between a causal variable and an outcomes" (Wilkinson et al., 1999, n.p.).

Lack of Literature Reviews

I predicted that about 50% of articles sampled in the current review would lack a literature review section. However, I am not confident about making a strong claim about the presence or absence of literature reviews in the articles in the current review because of the low levels of interrater agreement on this variable and on the other variables dealing with report elements. However, I think that the fact that two raters could not reliably agree on the presence or absence of key report elements; such as the literature review, research questions, report elements, description of participants, description of procedure; at least points out that these elements need to be explained more clearly. For example, if two raters cannot agree on whether or not there is a literature review in an academic paper, I am inclined to believe that the literature review is flawed in some way.

Assuming that the literature reviews in computer science education research articles are indeed lacking, then it is no surprise that the ACM SIGCSE Working Group on Challenges to Computer Science Education concluded that there is a lack of accumulated evidence and a tendency for computer science educators to "reinvent the wheel" (Almstrum et al., 2005, p. 191). Besides allowing evidence to accumulate and not reinventing the wheel, conducting thorough literature reviews takes some of the burden off researchers who are attempting to gather evidence for a claim since "good prior evidence often reduces the quality needed for later evidence" (Mark, Henry, & Julnes, 2000, p. 87).

Also, one conclusion that can be drawn from the fact that the literature review and other report elements variables had such low reliabilities is that the traditions of reporting differ significantly between what is suggested by the American Psychological suggestion and how most computer science education reports are structured. While not having agreed upon structures enables alternative styles of reporting to flourish and gives authors plenty of leeway to present their results, it makes it difficult for the reader to quickly extract needed information from the articles. Additionally, I hypothesize that the lack of agreed upon structures for computer science education articles leads to the omission of critical information needed in reports of research with human participants, such as a description of procedures and participants, especially by beginning researchers. Note that the report element variables; such as the lack of a literature review, the lack of information about participants or procedures, etc.; only pertained to articles that reported on investigations with human participants and not to other types of articles, such as program descriptions ortheoretical papers, in which the report structures would obviously differ from a report of an investigation with human participants.

Statistical Practices

The American Psychological Association (2001, p. 23) suggests that certain information be provided when certain statistical analyses are used. For example when parametric tests of location are used "a set of sufficient statistics consists of cell means, cell sample sizes, and some measures of variability. . . . Alternately, a set of sufficient statistics consists of cell means, along with the mean square error and degrees of freedom associated with the effect being tested." Second, the American Psychological Association (2001) and the American Psychological Association's Task Force on Statistical Inference Testing (Wilkinson et al., 1999) argue that it is best practice to report an effect size in addition to p-values. The results of this review showed that inferential analyses are conducted in 36% of cases when quantitative results are reported. When computer science educators do conduct inferential analyses, only a moderate proportion report informationally adequate statistics. Areas of concern include reporting a measure of centrality and dispersion for parametric analyses, reporting sample sizes and correlation or covariance matrices for correlational analyses, and summarizing raw data when nonparametric analyses are used.

Islands of Practice

In this section I discuss where there were or were not differences in research practices—in journals and conference proceedings, across regions, and across years. I used two different kinds of statistical approaches– χ2 analyses of crosstabulation and logistic regression–in my search for islands of practice. Most of the time those two approaches yielded the same results, sometimes they did not. In the cases where there was a discrepancy, I provide an explanation in this section. A summary of findings about islands of is provided in the list below:

1. There were no difference between journals and conference proceedings in terms of the proportions of anecdotal-only articles, explanatory descriptive articles, attitudes-only articles, and one-group posttest-only articles. Controlling for a region by forum type interaction, there is some evidence that the proportion of experimental/quasi-experimental articles is greater in conferences than in journals.

2. Region was a statistically significant predictor on every outcome variable except the proportion of one-group posttest-only articles.
   a. Controlling for other factors, North American articles had a higher proportion of anecdotal only articles than most other regions.
   b. North American articles had higher proportion of experimental/quasi-experimental articles than other regions.
   c. Middle Eastern articles had a much higher proportion of explanatory descriptive articles than articles from any other region.
   d. Asian-Pacific/Eurasian articles had a higher proportion of attitudes-only articles than did articles from other regions.

3. The proportion of anecdotal-only articles had decreased each year; the strongest decrease was seen in North American articles. Also, the proportion of explanatory descriptive articles had decreased every year.

Journal Vs. Conference Papers

There has been an ongoing debate in the field of computer science education about the relative merit that should be afforded to papers published in peer-reviewed journals and those published in peer-reviewed conference proceedings (see Frailey, 2006; Hodas, 2002). The outcomes of the debate about which academic publishing forums have the most merit are important to several groups. According to Walstrom, Hardgrave and Wilson (1995), those groups are:

- Selection, promotion, and tenure committees as they seek to secure and retain the best possible individuals for the faculty;
- Researchers as they seek to determine appropriate outlets for their research findings;
- Individuals seeking to identify the significant research streams in an academic discipline;
- Journal editors and associates as they seek to raise the quality of their journal [or conference] to the highest level possible;
- The academic discipline in question as it seeks to gain an identity of its own, especially as it relates to a young field;
- Students of the discipline as they seek to gain an understanding of what the discipline encompasses; and
- Librarians as they seek to wisely invest their ever-decreasing funds. (p. 93)

Particularly, the outcomes of the merit debate have serious economic consequences for academic professionals who work in a "publish-or-perish" environment. For example, Gill reports that "a published MIS [management information systems] referred journal article can be worth approximately $20,000 in incremental pay, over an assumed five-year lifetime, to a faculty member" (2001, p. 14). In the computing sciences, the relative academic worth afforded to journal and conference papers differs significantly from department to department. Some departments reportedly do not accept conference proceedings in the tenure review process (Hodas, 2002), Grudin (2004) reported that "some departments equate two conference papers to a journal article, or even award stature to papers in conferences that accept fewer than 25% of submissions" (p. 12), while others assign value to each article, whether published journal or conference proceedings, on a case-by-case basis (National Research Council, 1994). At any rate, the prevailing perception is that, generally, articles published in archival journals receive more academic merit than articles published in conference proceedings (National Research Council, 1994). Research conducted by the National Research Council has shown that researchers and university administrators who believe that journals are superior to conference proceedings believe so because of "the more critical reviewing and permanent record of the former" (p. 138).

There has been much research done in the field of MIS on the relative qualities of the different journal publication forums. The authors of that research (e.g., Katerattanakul, Han, & Hong, 2003; Rainer & Miller, 2005; Walstrom et al., 1995) generally took a citation analysis approach or measured the perceptions of those articles. However, that body of research is not directly applicable to this methodological review because they compared journals with journals and they conducted the study in the field of MIS, not computer science education. There are a few methodological reviews of the computer science education literature that have been published (Randolph, Bednarik, & Myller, 2006; Valentine, 2004). However, none of them specifically compared the methodological properties of journal and conference articles. However, one study that did compare journal articles with conference proceedings articles was conducted by the National Research Council (1994). In that study they compared computer science journals and conference publications on three variables: (a) time to publication, (b) median age of a reference, and (c) acceptance rate. The National Research Council's findings are listed below:

1. The median time from initial submission to publication in conference proceedings was 7 months while in journals it was 31 months.
2. The median age of a reference (the median difference between the date of an article's publication and the date of publication of the articles that were cited) was 3 years for conference proceeding articles and nearly 5 years for journal articles.
3. The acceptance rate for prestigious conference proceedings, which ranged from 18 to 23%, was slightly lower that the estimated acceptance rate for journals, 25 to 30%.

Although the National Research Council study (1994) provided some interesting results, it did not measure any construct dealing with the quality of the articles published in each of those forums. Given that the National Research Council's findings above are true, journal and conference articles might still differ substantially in terms of the quality of methodological practices used, which is one claim made by those who support giving more merit to journals.

If the variables–proportion of anecdotal-only articles, proportion of attitudes-only articles, proportions of articles using a one-group posttest-only design only, and proportion of experimental articles—are valid indicators of the methodological quality of articles, the hypothesis that computer science education journal articles are more methodologically sound than computer science education conference proceedings articles turned out to be wrong. In fact, there is some evidence that conference proceedings have a higher proportion of experimental/quasi-experimental articles than journal articles, when a region by forum type interaction is controlled for. Crosstabulations for Tables 39 through 43 showed that there were no statistically differences on any of the outcome variables, including the proportion of experimental/ quasi-experimental articles. When aggregating across regions and year, there is even a slightly greater proportion of experimental/quasi-experimental journal articles than conference articles (69.7% vs. 68.3%), see Table 40. However, using the logistic regression approach in which the unique effect of each predictor could be estimated and interactions could be modeled, there is evidence that the odds of a conference article's being experimental/quasi-experimental is greater than the odds for a journal paper. There was a statistically significant interaction between forum type and region. This interaction helps explain the incongruence between the aggregate, crosstabulation analysis and the logistic regression analysis.

Figure 9 shows that the proportion of experimental/quasi-experimental journal articles is much lower than the proportion of experimental/quasi-experimental conference papers for European and Asian-Pacific/Eurasian articles. However, the opposite is the case for North American articles; there are more experimental/quasi-experimental conference papers than there are experimental/quasi-experimental journal articles. My hypothesis for why this interaction exists rests on two assumptions.

The first is that journals are less influenced by regional affects than are conference proceedings. For example, authors who have a paper accepted at a conference are physically expected to appear at the conference to present their results. The effect is that people tend to attend, and submit papers to, conferences that are nearby. A quick glance at the conference proceedings included in this sample will support this point. Therefore, the research practices in a certain region will be reflected to some degree in the conference proceedings. The same does not hold for journals or holds to a lesser degree; authors of journal manuscripts are not expected to travel to the physical location where a journal is published.

The second assumption is that North American researchers tend to write and get published experimental/quasi-experimental articles more than European and Asian- Pacific/Eurasian et al. authors. This assumption is backed up from the region section of Table 57 and from Table 49. Therefore, because of the greater effect of region on conference proceedings than on journals and because of the tendency of North American researchers to do experimental research, the interaction is not surprising. The interaction seems to be strong enough that when included in the regression equation, it can switch the direction of the odds ratio (i.e., the predicted odds of a conference article's being an experimental/quasiexperimental article becomes greater than the odds of a journal article's being an experimental/quasi-experimental article.) Whether the interaction term is included or not, the results overall indicate that there are nonsignificant differences, or differences slightly in favor of conferences, in terms of the proportion of experimental/quasi-experimental articles in journals and conference proceedings. The results from both analyses indicate that there are no statistically significant differences between journals and conference proceedings in terms of the proportions of anecdotal-only, explanatory descriptive, attitudes-only, or one-group posttest-only articles. One limitation regarding this finding was that the coders were aware of whether the article being coded came from a conference proceeding or from a journal. Thus, it is plausible that experimenter bias could have come into play—the coders might have tended to code journal articles more leniently than conference articles because of a preexisting belief that journal articles are more methodologically sound. Blind review was not possible in this case because the length of the article would usually entail its status; if the article was five pages or less, it was most likely a conference proceeding paper. However, there is one reason that I believe that experimenter bias was not a serious threat in this study. If there had been experimenter bias, it should have worked in favor of the hypothesis that journal articles are more methodologically sound than conference proceedings articles; however, that was not the case.

In terms of informing policy for the personnel evaluation of computer science education researchers, the major implication of this finding is that it is inadvisable to summarily give less academic merit to conference proceedings than to journal articles, because their methodological soundness has been shown to be similar. I acknowledge, however, that the methodological soundness of an article should not be the only way that an article is evaluated. In essence, I agree with the Patterson, Snyder, and Ullman, representatives of the Computing Research Association, who wrote:

> For the purposes of evaluating a faculty member for promotion or tenure, there are two critical objectives of an evaluation: (a) establish a connection between a faculty member's intellectual contribution and the benefits claimed for it, and (b) determine the magnitude and significance of the impact. Both aspects can be documented, but it is more complicated than simply counting archival publications. . . . Not all papers in high quality publications are of great significance, and high quality papers can appear in lower quality venues. Publication's indirect approach to assessing impact implies that it is useful, but not definitive. The primary direct means of assessing impact—to document items (a) and (b) above—is by letters of evaluation from peers. (1999, pp. A-B)

Although publication counting and using merit formulas (e.g., that two conference papers are worth one journal article) are easy evaluation strategies, there can be no substitute for case-by-case assessment in which a variety of factors are taken into account in the gestalt of a faculty member's academic output.

Yearly Trends

Valentine (2004) identified several encouraging trends in computer science education research from 1984 to 1999. First, the number of technical symposium proceedings had been increasing each year. Second, the percentage of experimental articles (loosely defined as the author having made "any attempt at assessing the 'treatment' with some scientific analysis" [p. 256]) had increased since the mid '90s. Third, the percentage of Marco Polo articles (which probably would correspond with what I called anecdotal-only articles) had shown a yearly decrease. The findings of this methodological review show that two out of the three trends identified by Valentine (2004), from 1984 to 1999, continued in the years from 2000 to 2005. First, as is evident from Table 5, the number of articles in the SIGCSE Technical Symposium (and in computer science education forums in general) has still been on the rise. Second, the decline in the number of anecdotal-only/Marco Polo articles had continued to decline in the years from 2000-2005. The decline was most pronounced for North American articles. In contrast to what Valentine found, it was not found that the proportions of experimental articles had continued to increase into the years from 2000 to 2005. However, it is important to note here that I used a more conservative definition of experimental than did Valentine. I assume that, in addition to true experiments or quasiexperiments, Valentine would have included explanatory descriptive, exploratory descriptive, correlational, and causal comparative investigations in the "experimental" category. I, on the other hand, only included actual experiments or quasi-experiments in the experimental category.

Region of Origin

Concerning region of first author's origin, both the crosstabulation approach and the logistic regression approach revealed several differences in the way that computer science education researchers from institutions in different regions conduct research:

1. Computer science education researchers from North American institutions tended to do experimental research, while their European and Middle Eastern counterparts tended to not do experimental research;
2. Computer science education researchers from Middle Eastern institutions strongly tended to do explanatory descriptive (qualitative) research;
3. North American researchers tended to do anecdotal-only research more than their peers in other regions, but the proportions of North American anecdotal research articles had been on the decline while the proportions had been stable across time for the other regions; and
4. Computer science education researchers from Asian-Pacific or Eurasian institutions tended to measure attitudes only.

Disentangling the relationship between the factors related to the environment that a group of scientists work in and how they carry out their research is difficult (see Depaepe, 2002). It is like speculating how the work of the Vienna School, for example, would have been different had they been the Toledo (Ohio) School instead. Nonetheless, below I describe some of my hypotheses, which might be used to inform further investigations, about why the results may have turned out as they did. One possible reason for the tendency for North American education researchers to do experiments could be that the worth attributed to randomized field trials by the U.S. Department of Education, a major source of funding for U.S. education researchers, has something to do with the tendency of North American researchers (of whom most are from U.S. institutions) to do experimental research. The U.S. Department of Education (2002) made the following statement about the relative importance they give to descriptive studies and to "rigorous field trials of specific interventions":

> Descriptive implementation studies play a crucial role in understanding the impact of policy changes, but they are no substitute for rigorous field trials of specific interventions. Even with high-quality fast-response surveys, annual performance data, and descriptive studies, we still cannot answer the question on the minds of practitioners: "What works?" To be able to make causal links between interventions and outcomes, we need rigorous field trials, complete with random assignment, value-added analysis of longitudinal achievement data, and distinct interventions to study. This approach might be considered "research" rather than "evaluation." Whatever the name, the Department's evaluation agenda would be incomplete without it. It is a fair use of evaluation dollars because federal program funds are paying for the interventions to be studied. (Para. 24-26)

This policy is a hotly-debated topic in U.S. research and evaluation circles (see Donaldson & Christie, 2005; Julnes & Rog, in press; or Lawrenz & Huffman, 2006). Regardless of the propriety of this policy, the quote above shows that U.S. educational policymakers give value and funding priority to true experiments, and, it is not surprising then that many U.S. education researchers strive to do experimental research. Second, the tendency of European researchers to not do experimental research is congruent with the contemporary European decline in the popularity of the study of quantitative research methods. Rautopuro and Väisänen (2005); well-known Finnish, quantitative-research-method educators; wrote the following about the state of quantitative research methods, at least in Finland:

> The level of skills in the quantitative methods seems to be worrying. In educational science, too, the level of method used as well as how they are used in quantitative research in all levels—from master theses to dissertations—is getting out of hand. The students do not get excited of taking voluntary quantitative research methods courses and therefore are not capable to use them in their own research. Compulsory statistics courses, as well, are only a

necessity for the students and sometimes for the researcher, too. Moreover, one generation of educational researchers, at least partially, have lost the competence of applying quantitative research methods and because of this they have also lost the possibility to pass on the tradition of the use of these methods. (p. 273)

If Rautopuro and Väisänen's (2005) findings generalize to the rest of Europe (and there is reason to believe that it does — see European Science Foundation, 2004), then it is no surprise that there is a tendency for European computer science researchers to not do experimental research. One possible reason for this could be that the resurgence of the qualitative research tradition has had a greater influence in Europe than in North America, according to Fielding (2005). Fielding speculated that the "American uantitative approach was influential during this period [i.e., the resurgence of the qualitative method since the publication of Glaser and Strauss's Discover of Grounded Theory in 1967, Strauss and Corbin's revision of it in 1990, and Turner's influential 1981 paper on qualitative data analysis] too but qualitative methodology was arguably more secure in the European curriculum due to the import of hermeneutics in German social philosophy and the life history method in French and Italian sociology" (2005, para. 12).

Fielding (2005) also mentioned that qualitative research has become increasingly legitimized and institutionalized in the European social science research curriculum since the 1980s. One example of this institutionalization of qualitative research that Fielding provides are the postgraduate training guidelines written by the United Kingdom's Economic and Social Research Council (ESRC). According to Fielding those curriculum guidelines

> strongly emphasize qualitative methods and require that students understand archival, documentary and historical data, life stories, visual images and materials, ethnographic methods, cases studies and group discussions, at least one qualitative software package, and a range of analytic techniques including conversation analysis and discourse analysis. Since the guidelines are written by senior academics, they clearly index the institutionalization of qualitative methods. (Para. 21)

Concerning the finding that computer science education researchers affiliated with Middle Eastern institutions tended to do explanatory descriptive research, a quick examination of the Middle Eastern institutions from which the Middle Eastern articles came sheds light on this finding. Three Israeli institutions accounted for over half of the Middle Eastern computer science education articles. Those institutions were the Technion – Israel Institute of Technology, the Weizmann Institute of Science, and Tel-Aviv University, which contributed 23.1, 23.1, and 11.5% of the total number of Middle Eastern computer science articles included in this sample. One interesting finding was that North American papers had a significantly higher proportion of anecdotal-only papers than other regions (see Figure 7), but that this proportion had been declining over time in North American papers. As Figure 6 shows, in 2000 the proportion of North American anecdotal-only papers was about 80%; in 2005 the proportion was about equal with the proportions of other regions at about 30%.

Although I do not have any informed hypotheses about why the proportion of anecdotal-only North American papers would have been so much higher than in other regions in 2000, I do have one hypothesis about why the proportion of anecdotal-only articles had been declining steadily only in North America, besides the fact that extreme scores tend to regress towards the mean.

Given that more than one third of the total computer science articles came from the SIGCSE Conference Proceedings, which were held in the United States from 2000 through 2005, one possible explanation is that the decline in North American conference papers is heavily correlated with a decline in anecdotal-only papers in SIGCSE conference proceedings. (In fact, the Spearman correlation of the percent anecdotal-only by year between the SIGCSE Conference Proceedings and North American articles in general was quite high, $r(6) = .87$, p < .02.)

In addition, that decline in the proportion of anecdotal-only SIGCSE conference papers could be a result of the increased interest in the methodological qualities of the articles published in SIGCSE Proceedings, which is evident in recent SIGCSE Conference Proceedings articles, such as Valentine (2004), and working group reports, such as Almstrum, Ginat, Hazzan, and Clement (2003) and Almstrum and colleagues (2005). One flaw with this hypothesis though is that there has also been a recent interest in the methodological quality of computer science education research articles across the range of computer science publication forums, which is evident in articles such as Almstrum et al. (2002); Bouvier, Lewandowski, and Scott (2003); Carbone and Kaasbøøll (1998); Clear (2001); Daniels, Petre, and Berglund (1998); Fincher et al. (2005); Fincher and Petre (2004); Greening 1997); Lister (2005); Pears and colleagues (2005); Pears, Daniels, and Berglund (2002); Randolph, Bednarik, and Myller (2005), and Sandström and Daniels (2000), among others.

Differences Across Fields

Earlier I predicted that computer science education research would have the greatest proportion of papers that do not empirically deal with human participants, educational technology papers would have fewer of those papers than computer science education papers, and that education research proper papers would have the fewest of those types of papers. That prediction turned out to be correct. Assuming that the proportion of papers that do not empirically deal with human participants are, more or less, indicators of engineering and/or formalist traditions lingering in computer science education, then, it can be said that computer science education is a field in which the traditions of computer science research proper, especially the engineering tradition, bleed through to the practice of computer science education research. Computer science education researchers, as a whole, publish more "I engineered this intervention to certain specifications" types of articles and less "I empirically evaluated the effects of this intervention on student learning" types of articles than their counterparts in educational technology. In turn, educational technologists, as a whole, publish more engineering types of articles and less empirical types of articles than their counterparts in educational research proper.

In terms of the proportions of qualitative, quantitative, and mixed-methods research, computer science educators tended to use quantitative methods more frequently and qualitative research less frequently than their counterpart researchers in educational technology or education proper. This might come as a source of concern to the factions of computer science education researchers who call for more qualitative research, such as Ben-Ari, Berglund, Booth, and Holmboe (2004); Berglund, Daniels, and Pears (2006); Hazzan, Dubinsky, Eidelman, Sakhnini, and Teif (2006) and Lister (2003).

Profile of the Average Computer Science Education Paper

From these results, it is possible to create a profile of the average computer science education research paper. It is important to note that this profile is a synthesis of averages; there might not actually be an average paper that has this exact profile. Nonetheless, I include the average profile here because of the narrative efficiency in which it can characterize what computer science education research papers, in general, are like. The profile follows: The typical computer science education research paper is a 5-page conference paper written by two authors. The first author is most likely affiliated with a university in North America. If the article does not deal with human participants, then it is likely to be a description of some kind of an intervention, such as a new tool or a new way to teach a course. If the article does deal with human participants, then there is a 40% chance that it is basically a description of an intervention in which only anecdotal evidence is provided.

If more than anecdotal evidence is provided the authors probably used a one-group posttest-only design in which they gave out an attitude questionnaire, after the intervention was implemented, to a convenience sample of first-year undergraduate computer science students. The students were expected to report on how well they liked the intervention or how well they thought that the intervention helped them learn. Most likely, the authors presented raw statistics on the proportions of students who held particular attitudes.

Recommendations

●————————————————————————————————————●

In this section I report on what I consider to be the most important evidence-based recommendations for improving the current state of computer science education. Because I expect that the improvements will be most likely effected by editors and reviewers raising the bar in terms of the methodological quality of papers that get accepted for publication, I direct these recommendations primarily to the editors and reviewers of computer science education research forums. Also, these recommendations are relevant to funders of computer science research; to consumers of computer science education research, such as educational administrators; and, of course, to computer science education researchers themselves.

Accept Anecdotal Experience as a Means of Hypothesis Generation, But Not as the Sole Means of Hypothesis Confirmation

While a field probably cannot be built entirely on anecdotal experience (although some might not agree), that does not mean that anecdotal experience does not have an important role in scientific inquiry—it has an important role in the generation of hypotheses. Sometimes it is through anecdotal experience that researchers come to formulate important hypotheses. However, because of its informality, anecdotal experience is certainly a dubious type of evidence for hypothesis confirmation. Not accepting anecdotal evidence as a means of hypothesis confirmation is not to say that a human cannot make valid and reliable observations. However, there is a significant difference between a researcher reporting that "we noticed that students learned a lot from our program" and a researcher who reports on the results of a well-planned qualitative inquiry or on the results of carefully controlled direct observations of student behavior, for example. Also when anecdotal evidence is presented either as a rationale for a hypothesis to be investigated or as evidence to confirm a hypothesis, it should be clearly stated that anecdotal experience was the basis for that evidence.

Be Wary of Investigations That Only Measure Students' Self-Reports of Learning

Of course, stakeholders' reports about how much they have learned are important; however, it probably is not the only dependent of variable of interest in an educational intervention. As a measure of learning, as Guzdzial (in Almstrum et al., 2005) has pointed out, students' opinions are poor indicators of how much learning has actually occurred.

Insist That Authors Provide Some Kind of Information about the Reliability and Validity of Measures That They Use

Wilkinson et al. (1999) provided valuable advice to editors concerning this issue, especially in "a new and rapidly growing research area" (like computer science education). They advised,

> Editors and reviewers should pay special attention to the psychometric properties of the instrument used, and they might want to encourage revisions (even if not by the scale's author) to prevent the accumulation of results based on relatively invalid or unreliable measures. (n.p.)

Realize That the One-Group Posttest-Only Research Design Is Susceptible to Almost All Threats to Internal Validity

In the one-group posttest-only design, almost any influence could have caused the result. For example, in a one-group posttest-only design, if the independent variable was an automated tool to teach programming concepts and the dependent variable was the mastery of programming concepts, it is entirely possible that, for example, students already knew the concepts before using the tools, or that something other than the tool (e.g., the instructor) caused the mastery of the concepts. Experimental research designs that compare a factual to a counterfactual condition are much better at establishing causality than research designs that do not.

Report Informationally Adequate Statistics

When inferential statistics are used, be sure that the author includes enough information for the reader to understand the analysis used and to examine alternative hypotheses for the results that were found. The American Psychological Association (2001) gives the following guidelines:

> Because analytic technique depends on different aspects of the data, it is impossible to specify what constitutes a set of minimally adequate statistics for every analysis. However, a minimally adequate set usually includes at least the following: the per-cell sample size, the observed cell means (or frequencies of cases in each category for a categorical variable), the cell standard deviations, and an estimate of pooled within-cell variance. In the case of

multivariable analytic systems such as multivariate analyses, regression analyses, and structural equation modeling analyses, the mean(s), sample size(s), and the variance-covariance (or correlation) matrix or matrices are a part of a minimally adequate set of statistics. (p. 23)

Insist that Authors Provide Sufficient Detail
About Participants and Procedures

When authors report research on human participants be sure that they include adequate information about the participants, apparatus, and procedure. In terms of adequately describing participants the American Psychological Association (2001) suggests the following:

> When humans participated as the subjects of the study, report the procedures for selecting and assigning them and the agreements and payments made. . . . Report major demographic characteristics such as sex, age, and race/ethnicity, and where possible and appropriate, characteristics such as socio-economic status, disability status, and sexual orientation. When a particular demographic characteristic is an experimental variable or is important for the interpretation of results, describe the group specifically—for example, in terms of national origin, level of education, health status, and language preference . . . . Even when a characteristic is not an analytic variable, reporting it may give readers a more complete understanding of the sample and often proves useful in meta-analytic studies that incorporate the article's results. (pp. 18-19)

In terms of the adequate level of detail for the Procedures section, the American Psychological (2001) gives the following advice:

> The subsection on procedures summarizes each step in the execution of the research. Include the instructions to the participants, the formation of the groups, and the specific experimental manipulations. Describe randomization, counterbalancing, and other control features in the design. Summarize or paraphrase instructions, unless they are unusual or compose an experimental manipulation, in which case they may be presented verbatim. Most readers are familiar with standard testing procedures; unless new or unique procedures are used, do not describe them in detail.
>
> If a language other than English is used in the collection of information, the language should be specified. When an instrument is translated into another language, the specific method of translation should be described (e.g., back translation, in which a text is translated into another language and then back into the first to ensure that it is equivalent enough that the results can be compared.)
>
> Remember that the Method section should tell the reader what you did and how you did it in sufficient detail so that a reader could reasonably replicate your study. Methodological articles may defer highly detailed accounts of approaches (e.g., derivations and details of data simulation approaches) to an appendix. (p. 20)

In short, enough information should be provided about participants so that readers can determine generalization parameters and enough information should be provided about the procedure that it could be independently replicated.

An Example of a High-Quality Computer Science Education Research Article

In this section I examine in detail one article that I think is a particularly good example of high quality computer science education research and evaluate it in terms of the recommendations that I mentioned above. All though there were many high-quality articles in the sample that would have worked for this purpose, I chose Sajaniemi and Kuittinen's (2005) "An Experiment on Using Roles of Variables in Teaching Introductory Programming" because it was particularly clear and well-written and is exemplary in the areas that my recommendations relate to. (Although Jorma Sajaniemi works in the same department as I, this did not influence my choosing this article—at least that I am aware of. It was a random chance that this article was included in my sample in the first place.) The article is somewhat atypical in that that it is a 25-page journal paper (published in Computer Science Education), whereas most computer science education research papers are 5-page conference papers. To get a sense of what the article is about in general I have included the text from entire abstract below:

> Roles of variables is a new concept that captures tacit expert knowledge in a form that can be taught in introductory programming courses. A role describes some stereotypic use of variables, and only ten roles are needed to cover 99% of all variables in novice-level programs. This paper presents the results of an experiment where roles were introduced to novices learning Pascal programming. Students were divided into three groups that were instructed differently: in the traditional way with no treatment of roles; using roles throughout the course; and using a role-based program animator in addition to using roles in teaching. The results show that students are not only able to understand the role concept and to apply it in new situations but—more importantly—that roles provide students a new conceptual framework that enables them to mentally process program information in a way demonstrating good programming skills. Moreover, the use of the animator seems to foster the adoption of role knowledge. (p. 59)

According to the Publication Manual of the American Psychological Association (American Psychological Association, 2001) the abstract of an empirical report should describe

- the problem under investigation, in one sentence if possible;
- the participants or subjects, specifying pertinent characteristic, such as number, type, age, sex,...;
- the experimental method, including the apparatus, data-gathering procedures, [and] complete test names....;

- the findings, including statistical significance levels; and the conclusions and the implications or applications. (p. 14)

Sajaniemi and Kuitten's abstract described most of the information that the Publication Manual of the American Psychological Association calls for. The exceptions were, however, that Sajaniemi and Kuitten did not include as detailed information about participants as called for by the American Psychological Association, information about data-gathering procedures, and information about the significance level of findings. Overall, however, the abstract accurately summarizes the important parts of the article and, admittedly, Sajaniemi and Kuitten may have written their article according to some other publication manual than the Publication Manual of the American Psychological Association.

The introduction of their article clearly introduced the problem (a need for and lack of research on the role concept in teaching programming) and answered the following questions (from American Psychological Association, 2001, pp. 15-16):

1. Why is the problem important? (The answer could inform the teaching of programming.)
2. How do the hypothesis and the experimental design relate to the problem? (The hypothesis relates to a new way of teaching programming; the experimental design allows for an examination of the effects of that way of teaching programming or learning of programming.)
3. What are the theoretical implications of the study, and how does the study relate to previous literature? (The study informs theories about the different theories of teaching programming and can also inform other learning theories, such as the dual-coding theory, the cognitive constructivism theory, and the epistemic fidelity theory; the study relates to a new category of research on teaching of programming—software design patterns and roles of variables.)
4. What theoretical propositions are tested, and how were they derived. (The study tests the proposition that teaching roles of variables facilitates student learning of programming; Sajaniemi and Kuittinen provide a detailed research history of how those theoretical propositions were derived from previous research over the past 20 years.)

In the introduction of their article, Sajaniemi and Kuittien developed the background of the study with a discussion of the previous literature on teaching of programming, discussed how the theory being tested was derived, and gave a history and description of the intervention(s) that were used. As the Publication Manual of the American Psychological Association suggests, they cited "only works pertinent to the specific issue and not works of only tangential or general significance" (American Psychological Association, 2001, p. 16). Also, Sajaniemi and Kuitten clearly stated the purpose of their study, "to find out the effects of using the role concept in teaching programming to novices" (p. 60), and their research hypothesis—"introducing roles of variables in teaching facilitates learning to program" (p. 64).

The Publication Manual of the American Psychological Association (2001) suggests that the Method section should enable "the reader to evaluate the appropriateness of your methods and the reliability and validity of your results. It also permits experienced investigators to replicate the study if they so desire" (p. 17) and that it should, in most cases, contain the following subsections: participants, apparatus, and procedure. The Method section of Sajaniemi and Kuittinen's paper met all of those suggestions.

The Participants section of their paper (Sajaniemi and Kuittinen called it the Subjects section) provided detailed information about several participant variables that could have been confounded with treatment in the experiment. Some of those participants' variables were the number of subjects; gender; performance in high school mathematics, information technology, art; previous spreadsheet creation experience; previous programming courses; and previous programming experience. In short, they provided enough information about the participants that other researchers and practitioners would be able to establish generalization parameters and, by measuring variables that were thought to be possible confounding factors, were able to rule out a host of extraneous threats to internal validity.

In the Apparatus section, which Sajaniemi and Kuittinen labeled the "Materials" section, they provided detailed information on the measures that were used and even provided a web link, which actually worked, to the experimental materials that were used. The only information missing from the description of the examination was information about previous investigations on the validity or reliability of the measurement instrument (the examination).

In the beginning of the Method section and in the Procedure section Sajaniemi and Kuittinen provided copious detail about the research design (a between-subject design with the content of instruction as the between-subject factor, with researcher and grader blinding) and study procedures used. In my opinion, they provided enough information that other researchers could replicate the study.

In the Results section, Sajaniemi and Kuittenen did appropriate statistical analysis and presented informationally adequate statistics for the types of analyses the conducted–means, standard deviations, and n-sizes; correlational and raw effect sizes; and the value of the test statistic, degrees of freedom, and probability values. And they also presented a number of graphs to aid in the interpretation of results. The only information that would have improved this Results section is information on the interrater reliability estimates between graders.

In the Discussion section and Conclusion section, Sajaniemi and Kuittinen summarized their findings, revisited their research hypotheses, and related their findings back to the previous literature. They also outlined the implications of their study, discussed alternative hypotheses, and commented on study limitations.

This article can serve as a model for other computer science researchers in how to avoid the pitfalls common in the computer science research. First, they did a carefully controlled and rigorous study so that evidence could be collected that could help confirm or disconfirm their hypothesis. They used a design that is much better than the one-group posttest-only design for ruling out threats to internal validity. They created an instrument to measure learning instead of relying on students self-reports on whether they had learned or not. Although they did not provide information about the psychometric properties of their measurement instrument, they did describe the instrument in detail and their rationale for its validity. Also, they gave readers direct access to the actual measurement instrument that was used so that the readers could make their own judgments about the psychometric properties of the instrument. They provided rich enough detail of the participants, materials, and procedures used that the reader could clearly understand what happened in the experiment and could even replicate it. Finally, they provided informationally adequate statistics in the Results section. It is true that they had 25 pages in which to work and that normally computer science education research forums allow only up to 5 pages. Nevertheless, a 5-page empirical report should also have the same elements as a 25-page report–only the level of detail might change. Articles such as Clark, Anderson, and Chalmers (2002); Lee et al. (2002); and Olson et al. (2002), although in the field of medical science, are good examples of how empirical reports can be written in such a way that they are complete, but also very concise.

Chapter 5

———————————————————————————

Conclusion

In this line of research, I used a content analysis approach to conduct a methodological review of the articles published in mainstream computer science education forums from 2000 to 2005. Of the population of articles published during that time a random sample of 352 articles was drawn; each article was reviewed in terms of its general characteristics; the type of methods used; the research design used; the independent, dependent, and mediating or moderating variables used; the measures used; and statistical practices used. The major findings from the review are listed below:

1. About one third of articles did not report research on human participants.
2. Most of the articles that did not deal with human participants were program descriptions.
3. Nearly 40% of articles that dealt with human participants only provided anecdotal evidence for their claims.
4. Of the articles that provided more than anecdotal evidence, most articles used experimental/quasi-experimental or explanatory descriptive methods.
5. Of the articles that used an experimental research design, the majority used a one-group posttest-only design exclusively.
6. Student instruction, attitudes, and gender were the most frequent independent, dependent, and mediating/moderating variables, respectively.
7. Questionnaires were clearly the most frequently used type of measurement instrument. Almost all of the measurement instruments that should have psychometric information provided about them did not have psychometric information provided.
8. When inferential statistics were used, the amount of statistical information used was inadequate in many cases.
9. There was no difference in major methodological characteristics between articles published in computer science education journals and those published in peer-reviewed conference proceedings. However, there is some evidence that when controlling for the interaction between region and forum type, the odds of an article's being

———————————————————————————

experimental/quasi-experimental was higher in conference proceedings.

10. There was a decreasing yearly trend in the number of anecdotal-only articles and in the number of articles that used explanatory descriptive methods.

11. First authors affiliated with North American institutions tended to publish papers in which experimental/quasi-experimental papers were used; first authors affiliated with Middle Eastern or European institutions tended not to publish papers in which experimental or quasi-experimental methods were used.

12. First authors affiliated with Middle Eastern institutions strongly tended to publish explanatory descriptive articles.

13. First authors affiliated with Asian-Pacific or Eurasian institutions tended to publish articles in which attitudes were the sole independent variable.

14. First authors affiliated with North American institutions tended to publish anecdotal-only articles; however, that proportion of North American anecdotal-only articles had declined linearly over time and was about equal to the proportion in other regions by 2005.

15. Computer science education research forums published more engineering-oriented program-description types of papers than educational technology forums published and much more than education research proper forums published.

16. Computer science education researchers, in general, tended to use quantitative methods and tended not to use qualitative methods more than their counterparts in educational technology or education research proper.

Based on these findings, I made the following recommendations to editors, reviewers, authors, funders, and consumers of computer science education research:

1. Accept anecdotal experience as a means of hypothesis generation, but not as the sole means of hypothesis confirmation.

2. Be wary of investigations that measure only students' attitudes and self-reports of learning as a result of an intervention.

3. Insist that authors provide some kind of information about the reliability and validity of measures that they use.

4. Realize that the one-group posttest-only research design is susceptible to almost all threats to internal validity.

5. Encourage authors to report informationally adequate statistics.

6. Insist that authors provide sufficient detail about participants and procedures.

Computer Science Education Research at the Crossroads

Based on the results of this review, I can say that what computer science educators have so far been great at is generating a large number of informed research hypotheses, based on anecdotal experience or on poorly designed investigations. However, they have not systematically tested these hypotheses. This leaves computer science education at a crossroads. To the crossroads computer science education researchers bring a proliferation of well-informed hypotheses. What will happen to these hypotheses remains to be seen.

One option is that these informed hypotheses will overtime, through repeated exposure, "on the basis of 'success stories' and slick sales pitches" (Holloway, 1995, p. 20) come to be widely accepted as truths although having never been empirically verified. That is, they will become folk conclusions. (I use the term folk conclusions instead of folk theorems [see Harel, 1980] or folk myths [see Denning, 1980] since the validity of the conclusion has not yet been empirically determined.)

The consequences of accepting folk conclusions that are not actually true can be serious. Although speaking in the context of software engineering, but which probably still applies to some degree computing education as well, Holloway (1995) wrote:

> I pray that it will not take the loss of hundreds of lives in an airplane crash, or even the loss of millions of dollars in a financial system collapse, before we acknowledge our ignorance and redirect our efforts away from [promoting folk conclusions] and towards developing a valid epistemological foundation. (p. 21)

Because scientific knowledge usually develops cumulatively, if informed hypotheses are allowed to developed into folk conclusions, then layers of folk conclusions (both true and untrue) will become inexorably embedded in the cumulative knowledge of what is known about computer science education. Computer science education will become a field of research whose foundational knowledge is based on conclusions that are believed to be true, but which have never been empirically verified. Indeed, as Holloway suggests "resting an entire discipline on such a shaky epistemological foundation is absurd . . ." (1995, p. 21). In the same vein, basing the future of an entire discipline on such a shaky epistemological foundation is also absurd. I am not arguing, however, that hypothesis generation or any other type of research activity in computer science education should be abandoned altogether. There needs to be a requisite variety of methods to draw from so that a rich variety of research acts can be carried out. Also, hypothesis generation is inexorably tied with innovation.

What I am arguing is that the proportions of research methods being used needs to be congruent with the current challenges and problems in computer science education. If the ACM SIGCSE's Working Group on Challenges to Computer Science Education is correct that the current challenges involve a lack or rigor and accumulated evidence, then it makes sense to shift the balance from one that emphasizes anecdotal evidence and hypothesis generation to one that emphasizes rigorous methods and hypothesis confirmation. Coming back to the discussion of the crossroads, the sustainable path for computer science education involves building on the hypotheses of the past and striking a balance between innovation and experimentation in the future.

REFERENCES

Agresti, A. (1996). *An introduction to categorical data analysis*. New York: Wiley & Sons.

Alexander, S. & Hedberg, J. G. (1994). Evaluating technology-based learning: Which model? In K. Beattie, C. McNaught, & S. Willis (Eds.) *Multimedia in higher education: Designing for change in teaching and learning*. Amsterdam: Elsevier.

Almstrum, V. L., Ginat, D., Hazzan, O. & Clement, J. (2003). Transfer to/from computing science education: The case of science education research. In *Proceedings of the 34ᵗʰ SIGCSE Technical Symposium on Computer Science Education* (pp. 303-304). New York: ACM Press.

Almstrum, V. L., Ginat, D., Hazzan, O. & Morely, T. (2002). Import and export to/from computing science education. The case of mathematics education research. In *Proceedings of the 7ᵗʰ Annual Conference on Innovation and Technology in Computer Science Education ITiCSE '02* (pp. 193-194). New York: ACM Press.

Almstrum, V. L., Hazzan, O., Guzdzial, M. & Petre, M. (2005). Challenges to computer science education research. In *Proceedings of the 36ᵗʰ SIGCSE Technical Symposium on Computer Science Education SIGCSE '05* (pp. 191-192). New York: ACM Press.

American Psychological Association. (2001). *Publication manual of the American Psychological Association,* 5th ed. Washington, D.C.: American Psychological Association.

Babbitt, T. (2001). A tale of two shortages: An analysis of the IT professional and MIS faculty shortages. In *Proceedings of the 2001 ACM SIGCPR Conference on Computer Personnel Research* (pp. 21-23). New York: ACM Press.

Ben-Ari, M., Berglund, A., Booth, S., & Holmboe, C. (2004). What do we mean by theoretically sound research in computer science education? In *Proceedings of the 9ᵗʰ Annual Conference on Innovation and Technology in Computer Science Education ITiCSE '04* (pp. 230-231). New York: ACM Press.

Berglund, A., Daniels, M., & Pears, A. (2006). Qualitative research projects in computing education research. In *Proceedings of the 8ᵗʰ Australian Conference on Computing Education, 52,* (pp. 25-33). Darlinghurst, Australia: Australian Computer Society.

Bouvier, D., Lewandowski, G., & Scott, T. (2003). Developing a computer science education research program. *Journal of Computing Sciences in Colleges, 19*(1), 218.

Brennan, R. L. & Prediger, D. J. (1981). Coefficient kappa: Some uses, misuses, and alternatives. *Educational and Psychological Measurement, 41,* 687-699.

Campbell, D. T. (1975). "Degrees of freedom" and the case study. *Comparative Political Studies, 8,*

178-193.

Caffarella, E. P. (1999). The major themes and trends in doctoral dissertation research in educational technology from 1977 through 1998. In *Proceedings of selected research and development papers presented at the National Convention of the Association for Educational Communications and Technology (AECT)*. (ERIC Document Reproduction Service No. ED 436 178)

Carbone, A. & Kaasbøøll, J. (1998). A survey of methods used to evaluation computer science teaching. *SIGCSE Bulletin, 30*(3), 41-45

Clancy, M., Stasko, J., Guzdzial, M., Fincher, S., & Dale, N. (2001). Model and areas for CS education research. *Computer Science Education, 11*(4), 323-341.

Clark, M., Anderson, P., & Chalmers, I. (2002). Discussion sections in reports of controlled trials published in general medical journals. *Journal of the American Medical Association, 287*(21), 2799-2800.

Clark, R. E. & Snow, R. E. (1975). Alternative designs for instructional technology research. *AV Communications Review, 23*, 373- 394.

Clear, T. (2001). Thinking issues. Research paradigms and the nature and meaning of truth. *SIGCSE Bulletin, 33*(2), 9-10.

Cohen, J. (1994). The earth is round *(p < .05)*. *American Psychologist, 49*, 997-1003.

Cooper, H. M. (1988). Organizing knowledge synthesis: A taxonomy of literature reviews. *Knowledge in Society, 1*, 104-126.

Cooper, H. M. & Hedges, L. V. (1994). Research synthesis as a scientific enterprise. In H. Cooper and L. V. Hedges (Eds.), *The Handbook of Research Synthesis* (pp. 3- 14). New York: Russell Sage Foundation.

Daniels, M., Petre, M. & Berglund, A. (1998). Building a rigorous research agenda into changes to teaching. In *Proceedings of the 3rd Australasian Conference on Computer Science Education* (pp. 203-209). New York: ACM Press.

Denning, P. J. (1980). On folk theorems, and folk myths. *Communications of the ACM, 23*(9), 493-494.

Denning, P. J., Comer, D. E., Gries, D., Mulder, M. C., Tucker, A., Turner, A. J., & Young, P. R. (1989). Computing as a discipline. *Communications of the ACM, 32*(1), 9-23.

Depaepe, M. (2002). A comparative history of the educational sciences: The comparability of the incomparable? *European Education Research Journal, 1*(1), 118-122.

Dick, W. & Dick, W. D. (1985). Analytical and empirical comparisons of the *Journal of Instructional Development* and *Educational Communication and Technology Journal*. *Educational Technology Research & Development, 37*(1), 81-87.

Donaldson, S. I. & Christie, C. A. (2005). The 2004 Claremont Debate: Lipsey vs. Scriven. Determining causality in program evaluation and applied research: Should experimental evidence be the gold standard. *Journal of Multidisciplinary Evaluation, 3,* pp. 60-77.

Driscoll, M. P. & Dick, W. (1999). New research paradigms in instructional technology. An inquiry. *Educational Technology Research & Development, 47*(2), 7-18.

Edgington, E. S. (1964). A tabulation of inferential statistics used in psychology journals. *American Psychologist, 19*, 202-203.

Elmore, P. B. & Woehlke, P. J. (1988). Statistical methods employed in *American Educational Research Journal, Educational Researcher, and Review of Educational Research* from 1978 to 1987. *Educational Researcher, 17*(9), 19-20.

Elmore, P. B. & Woehlke, P. L. (1998, April). Twenty years of research methods employed in *American Education Research Journal, Educational Researcher*, and *Review of Educational Research.* Paper presented at the *Annual Meeting of the American Educational Research Association*, San Diego, CA.

Ely, D. (1999). Toward a philosophy of instructional technology. Thirty years on. *British Journal of Educational Technology, 30*(4), 305-310.

European Science Foundation. (2004, October). *Quantitative methods in the social sciences (QMSS): An ESF scientific programme.* Retrieved August 24, 2006 from http://www.esf.org/publication/193/QMSS.pdf

Fielding, N. (2005, May). The resurgence, legitmation and institutionalization of qualitative methods. *Forum Qualitative Sozialforschung / Forum: Qualitative Social Research, 6*(2), Article 32. Retrieved September 25, 2006 from http://www.qualitative-research.net/fqs-texte/2-05/05-2-32-e.htm

Fincher, S., Lister, R., Clear, T., Robins, A., Teneberg, J., & Petre, M. (2005). Multi-institutional, multi-national studies in CSEd research: Some design considerations and trade-offs. In *Proceedings of the 2005 International Workshop on Computing Education Research ICER '05* (pp. 111-121). New York: ACM Press.

Fincher, S. & Petre, M. (2004). *Computer Science Education Research.* London: Taylor & Francis.

Frailey, D. J. (2006, July 10). *What price publications?* Message posted to SIGCSE Member Forum electronic mailing list. Retrieved October 24, 2006 from http://listserv.acm.org/scripts/wa.exe?A2=ind0607b&L=sigcse-members&F=&S=&P=2865

Gall, M. D., Borg, W. R., & Gall, J. P. (1996). *Educational research: An introduction*, 6th ed. New York: Longman.

Garson, D. V. (2006). *Sampling*. Retrieved March 28, 2006 from North Carolina State University, College of Humanites & Social Science Web site: http://www2.chass.ncsu.edu/garson/PA765/sampling.htm

Gill, T. G. (2001). What's an MIS paper worth? (An exploratory analysis). *The DATA BASE for Advances in Information Systems, 32*(2), 14-33.

Glaser, B., & Strauss, A. (1967). *The discovery of grounded theory*. New York: Aldine.

Glass, R. L., Ramesh, V. & Vessey, I. (2004). An analysis of research in computing disciplines. *Communications of the ACM, 47*(6), 89-94.

Good, P. I. (2001). *Resampling methods. A practical guide to data analysis* (2nd ed.). Boston: Birkhäuser.

Goodwin, L. D. & Goodwin, W. L. (1985a). An analysis of the statistical techniques used in the *Journal of Educational Psychology*, 1979-1983. *Educational Psychologist, 20*, 13-21.

Goodwin, L. D. & Goodwin, W. L. (1985b). Statistical techniques in AERJ articles, 1979-1983: The preparation of graduate students to read the educational research literature. *Educational Researcher, 14*(2), 5-11.

Gorard, S. & Taylor, C. (2004). Combining methods in educational and social research. Berkshire, England: Open University Press.

Greening, T. (1997). Paradigms for educational research in computer science. In *Proceedings of the 2nd Australasian Conference on Computer Science Education* (pp. 47-51).

Grosberg, J. (n.d.). *Statistics 101* [Computer Software]. Retrieved July 11, 2006 from http://www.statistics101.net/index.htm

Grudin, J. (2004). Crossing the divide. *ACM Transactions on Computer-Human Interaction, 11*(1), 1-25.

Harel, D. (1980). On folk theorems. *Communications of the ACM, 23*(7), 379-494.

Hazzan, O., Dubinsky, Y., Eidelman, L., Sakhnini, V., & Teif, M. (2006). Qualitative research in computer science education. In *Proceedings of the 37th SIGCSE Technical Symposium on Computer Science Education* (pp. 408-412). New York: ACM Press.

Higgins, N., Sullivan, H., Harper-Marinick, M. & Lopez, C. (1999). Perspectives on educational technology research and development. *Educational Technology Research & Development, 37*(1), 7-17.

Hodas, J. S. (2002, November 15). *Re: Request for permission: appropriate journals (fwd)*. Message posted to SIGCSE Member Forum electronic mailing list. Retrieved October 24, 2006 from http://listserv.acm.org/scripts/wa.exe?A2=ind0211C&L=SIGCSE/MEMBERS&P=R3 50&I=-3

Holloway, C. M. (1995). Software engineering and epistemology. *Software Engineering Notes, 20*(2),

20-21.

Holmboe, C., McIver, L. & George, C. (2001). Research agenda for computer science. In G. Kododa (Ed.), *Proceedings of the 13th Annual Workshop of the Pscyhology of Programming Interest Group,* (pp. 207-223).

Huwiler-Müntener, K., Jüni, P., Junker, C., & Egger, M. (2002). Quality of reporting of randomized trials as a measure of methodologic quality. *Journal of the American Medical Association, 287*(21), 2801-2804.

Julnes, G. & Rog, D. J. (in press). Informing federal policies on evaluation methodology: Building the evidence for method choice in government sponsored evaluation. *New Directions for Evluation.*

Kalton, G. (1983). *Introduction to survey sampling.* London: Sage.

Katerattanakul, P., Han, B., and Hong, S. (2003). Objective quality ranking of computer journals. *Communications of the ACM, 46*(10), 111-114.

Keselman, H. J., Huberty, C. J., Lix, L. M., Olejnik, S., Cribbie, R., Donahue, B. Kowalkchuk, R R. K., Lowman, L. L., Petoskey, M. D., Keselman, J. C., & Levin, J. R. (1998). Statistical practices of educational researchers. An analysis of their ANOVA, MANOVA, and ANCOVA analyses. *Review of Educational Research, 68,* 350-86.

Kinnunen, P. (n.d.). *Guidelines of Computer Science Education Research.* Retrieved November 29, 2005 from http://www.cs.hut.fi/Research/COMPSER/ROLEP/seminaari-k05/S_05-nettiin/Guid elines_of_CSE-teksti-paivi.pdf

Kirk, R. E. (1996). Practical significance. A concept whose time has come. *Educational and Psychological Measurement, 56,* 746-759.

Kish, I. (1987). *Statistical design for research.* New York: Wiley.

Klein, J. D. (1997). *ETR&D* development: An analysis of content and survey of future direction. *Educational Technology Research & Development, 45*(3), 57-62

Lavori, P. W., Louis, T. A., Bailar, J. C., & Polansky, H. (1986). Design of experiments: Parallel comparisons of treatments. In J. C. Bailar & F. Mosteller (Eds.), *Medical uses of statistics* (pp. 61-82.) Waltham, MA: New England Journal of Medicine.

Lawrenz, F., & Huffman, D. (2006). Methodological Pluralism: The gold standard of STEM evaluation. In D. Huffman & F. Lawrenz (Eds.), New directions for evaluation: Critical issues in STEM evaluation (pp. 19-34). Hoboken, NJ: Wiley.

Lawrenz, F., Keiser, N., & Lavoir, B. (2003). Evaluative site visits: A methodological review. *American Journal of Evaluation, 24,* 341-352.

Lee, K. P., Schotland, M., Bachetti, P., & Bero, L. A. (2002). Association of journal quality indicators with methodological quality of clinical research articles. *Journal of the American Medical Association, 287*(1), 2805-2808.

Lister, R. (2003). A research manifesto, and the relevance of phenomenography. *ACM SIGCSE Bulletin, 35*(2), 15-16.

Lister, R. (2005). CS research: Mixed methods: Positivists are from Mars, constructivists are from Venus. *SIGCSE Bulletin, 37*(4), 18-19.

Mark, M. M., Henry, G. T., Julnes, G. (2000). *Evaluation: An integrated framework for understanding, guiding, and improving policies and programs.* San Francisco: Jossey-Bass.

Mohr, L. B. (1999). The qualitative method of impact analysis. *American Journal of Evaluation, 20*(1), 69-84.

National Research Council. (1994). *Academic careers for computer scientists and engineers.* National Academy Press, Washington, D.C.

National Research Council Committee on Information Technology Literacy. (1999, May). *Chapter 1: Why know about information technology: Being fluent with information technology.* Washington, DC: National Academy Press. Retrieved October 12, 2004 from http://books.nap.edu/html/beingfluent/ch1.html

Neuendorf, K. A. (2002) *The content analysis handbook.* Thousand Oaks, CA: Sage.

Olson, C. M., Rennie, D., Cook, D., Dickersin, K., Flanagin, A., Hogan, J. W., et al. (2002). Publication bias in editorial decision making. *Journal of the American Medical Association, 287*(21), 2825-2828.

Patterson, D., Snyder, L., & Ullman, J. (1999, September). Best practices memo: Evaluating computer scientists and engineers for promotion and tenure. *Computing Research New,* A-B.

Pears, A., Daniels, M. & Berglund, A. (2002). Describing computer science education research. An academic process view. In *Proceedings of the Conference on Simulation and Multimedia in Engineering Education, ICSEE 2002* (pp. 99-104).

Pears, A., Seidman, S., Eney, C., Kinnunen, P., & Malmi, L. (2005). ITiCSE 2005 working group reports: Constructing a core literature for computing education research. *SIGCSE Bulletin, 37*(4), 152-161.

Phipps. R, & Merisotis, J. (1999, April). *What's the difference: A review of the contemporary research on the effectiveness of distance learning in higher education.* The Institute for higher education policy. Retrieved July 27, 2006 from http://www.ihep.org/Pubs/PDF/Difference.pdf

Rainer, R. K. Jr. & Miller, M. D. (2005). Examining differences across journal rankings. *Communications of the ACM, 48*(2), 91-94.

Randolph, J. J. (2005) *A methodological review of the program evaluations in K-12 computer science education.* Manuscript submitted for publication. Available online http://www.geocities.com/justusrandolph/cse_eval_review.pdf

Randolph, J.J. (in press). What's the difference, still: A follow-up review of the quantitative research methodology in distance learning. *Informatics in Education.*

Randolph, J. J., Bednarik, R. & Myller, N. (2005). A methodological review of the articles published in the proceedings of Koli Calling 2001-2004. In T. Salakoski, T. Mäntylä, & M. Laakso (Eds.), *Proceedings of the 5th Annual Finnish / Baltic Sea Conference on Computer Science Education* (pp. 103-109). Finland: Helsinki University of Technology Press. Retrieved March 19, 2006 from http://www.it.utu.fi/koli05/proceedings/final_composition.b5.060207.pdf

Randolph, J. J., Bednarik, R., Silander, P., Lopez-Gonzalez, J., Myller, N., & Sutinen, E. (2005). A critical review of research methodologies reported in the full-papers of ICALT 2004. In *Proceedings of the Fifth International Conference on Advanced Learning Technologies* (pp. 10-14). Los Alamitos, CA: IEEE Press. Available online: http://ieeexplore.ieee.org/xpls/abs_all.jsp?isnumber=32317&arnumber=1508593&count=303&index=4

Randolph, J. J., & Hartikainen, E. (2005). A review of resources for K-12 computer-science-education program evaluation. *Yhtenäistyvät vai erilaistuvat oppimisen ja koulutuksen polut: Kasvatustieteen päivien 2004 verkkojulkaisu* [Electronic Proceedings of the Finnish Education Research Days Conference 2004] (pp. 183-193). Finland: University of Joensuu Press. Available online: http://www.geocities.com/justusrandolph/review_of_resources.pdf

Randolph, J.J., Hartikainen, E., & Kähkönen, E. (November, 2004). Lessons learned from developing a procedure for the critical review of educational technology research. Paper presented at *Kasvatustieteen Päivät 2004* [Finnish Education Research Days Conference 2004], Joensuu, Finland.

Ranis, S. H., & Walters, B. H. (2004). Education research as a contested enterprise: The deliberations of the SSRC-NAE Joint Committee on Education Research. *European Education Research Journal, 3*(4), 795-806.

Rautopuro, J. & Väisänen, P. (2005). DEEP WATER? Quantitative research methods in educational science in Finland. *In M-L Julkunen (Ed.), Learning and instruction in multiple contexts and settings III. Proceedings of the Fifth Joensuu Symposium on Learning and Instruction. Bulletins of the Faculty of Education, 96,* (pp. 273-293). Finland: University of Joensuu Press.

Reeves, T. C. (1995). *Questioning the questions of instructional technology research.* Retrieved October 19,

2004 from http://www.hbg.ps.edu/bsed/intro/docs/dean/

Resampling Stats (Version 5.0.2) [Computer software and manual]. (1999). Arlington, VA: Resampling Stats, Inc.

Research Assessment Exercise. (2001). Retrieved July 28, 206 from http://www.hero.ac.uk/rae/

Ross, S. M. & Morrison, G. R. (2004). Experimental research methods. In D. J. Jonassen (Ed.) *Handbook of research on educational communications and technology* (2nd ed., pp. 1021-1043.) Mahwah, NJ: Lawrence Earlbaum.

Sajaniemi, J., & Kuittinen, M. (2005). An experiment on using roles of variables in teaching introductory programming. *Computer Science Education, 15*(1), 59-82.

Sample Planning Wizard [Computer software]. (2005). Stat Trek.

Sandström, A. & Daniels, M. (2000). Time studies as a tool for (computer science) education research. In *Proceedings of the Australasian Conference on Computing Education* (pp. 208-214). New York: ACM Press.

Scriven, M. (1976). Maximizing the power of causal investigations: The modus operandi method. In G. V. Glass (Ed.), *Evaluation studies review annual,* Vol. 1 (pp. 101-118). Beverly Hills, CA: Sage Publications.

Shadish, W. R., Cook, T. D., & Campbell, D. T. (2002). *Experimental and quasi-experimental designs for generalized causal inference.* Boston: Houghton Mifflin.

Simon, J. L. (1997). *Resampling: The new statistics.* Arlington, VA: Resampling Stats, Inc.

Stevens, J. (1999). Intermediate statistics: A modern approach (2nd ed.). Mawwah, NJ: Lawrence Earlbaum and Associates.

Strauss, A., & Corbin, J. (1990). *Basics of qualitative research.* Newbury Park, CA: Sage.

Tedre, M. (2006). The development of computer science: A sociocultural perspective. Dissertation, University of Joensuu. Retrieved October 24, 2006 from http://joypub.joensuu.fi/publications/dissertations/tedre_development/index_en.html

Test, D. W., Fowler, C. H., Brewer, D. M., & Wood, W. M. (2005). A content and methodological review of self-advocacy intervention studies. *Exceptional Children, 72*, 101-125.

Thompson, B. & Snyder, P. A. (1998). Statistical significance and reliability analyses in recent JCD research articles. *Journal of Counseling and Development, 76*, 436- 441.

Tichy, W. F., Lukowicz, P., Prechelt, L., & Heinz, E. A. (1995). Experimental evaluation in computer science. A quantitative study. *Journal of Systems and Software, 28*, 9-18.

Tucker, A., Deck, F., Jones, J., McCowan, D., Stephenson, C., & Verno, A. (ACM K-12 Education Task Force Curriculum Committee). (2003, October 22). *A model curriculum for K-12 computer science: Final report of the ACM K-12 Education Task Force Curriculum Committee.* Retrieved

October 20, 2004, from http://www.acm.org/education/k12/k12final1022.pdf

Turner, B. (1981). Some practical aspects of qualitative data analysis. *Quality and Quantity, 15,* 225-247.

U.S. Department of Education. (2002). *Press release: New directions for program evaluation at the U.S. Department of Education.* Retrieved August 24, 2006 from http://www.ed.gov/news/pressreleases/2002/04/evaluation.html

U.S. Department of Labor - Bureau of Labor Statistics. (n.d.a). Computer support specialists and systems administrators. In *Occupational outlook handbook, 2004-2005 edition.* Retrieved October 22, 2004 from http://www.bls.gov/oco/pdf/oco/ocos268.pdf

U.S. Department of Labor - Bureau of Labor Statistics –. (n.d.b). Computer systems analysts database administrators, and computer scientists. In *Occupational outlook handbook, 2004-2005 edition.* Retrieved October 22, 2004 from http://www.bls.gov/oco/pdf/oco/ocos042.pdf

Valentine, D. W. (2004). CS educational research: A meta-analysis of SIGCSE technical symposium proceedings. In *Proceedings of the 35th Technical Symposium on Computer Science Education* (pp. 255-259). New York: ACM Press.

Walstrom, K. A., Hardgrave, B. G., & Wilson, R. L. (1995). Forums for management information systems scholars. *Communications of the ACM, 38*(3), 93-107.

West, C. K., Carmody, C. & Stallings, W. M. (1983). The quality of research articles in *Journal of Educational Research,* 1970 and 1980. *Journal of Educational Research, 77,* 70-76.

Wilkinson, L. & Task Force on Statistical Inference. (1999). Statistical methods in psychology journals: Guidelines and explanations [Electronic version]. *American Psychologist, 54,* 594-604.

Williamson, A., Nodder, C., Baker, P. (2001). *A critique of educational technology research in New Zealand since 1996.* Retrieved October 20, 2004 from http://www.ascilite.org.av/conferences/melbourne01/pdf/papers/williamson.pdf

Zelkowitz, M. V. & Wallace, D. (1997). Experimental validation in software engineering. *Information and Software Technology, 39,* 735-743.

APPENDICES

APENDIX A

A List of the Articles Included in the Sample

Abunawass, A., Lloyd, W., & Rudolph, E. (2004). COMPASS: A CS program assessment project. *ITiCSE '04: Proceedings of the 9th Annual SIGCSE Conference on Innovation and Technology in Computer Science Education,* Leeds, United Kingdom, 127-131. Retrieved September 4, 2006 from http://doi.acm.org/10.1145/1007996.1008031

Acharya, M., & Funderlic, R. (2003). 'Laurel and hardy' model for analyzing process synchronization algorithms and primitives. *SIGCSE Bulletin (Association for Computing Machinery, Special Interest Group on Computer Science Education), 35*(2), 107-110.

Adams, J. C., Bauer, V., & Baichoo, S. (2003). An expanding pipeline: Gender in mauritius. *SIGCSE '03: Proceedings of the 34th SIGCSE Technical Symposium on Computer Science Education,* Reno, NV, United States. 59-63, Retrieved September 4, 2006 from http://doi.acm.org/10.1145/611892.611932

Adams, J., Nevison, C., & Schaller, N. C. (2000). Parallel computing to start the millennium. *SIGCSE '00: Proceedings of the Thirty-First SIGCSE Technical Symposium on Computer Science Education,* Austin, TX, United States, 65-69. Retrieved September 4, 2006 from http://doi.acm.org/10.1145/330908.331815

Aharoni, D. (2000). Cogito, ergo sum! cognitive processes of students dealing with data structures. *SIGCSE '00: Proceedings of the Thirty-First SIGCSE Technical Symposium on Computer Science Education,* Austin, TX, United States, 26-30. Retrieved September 4, 2006 from http://doi.acm.org/10.1145/330908.331804

Aiken, R., Kock, N., & Mandviwalla, M. (2000). Fluency in information technology: A second course for non-CIS majors. *SIGCSE '00: Proceedings of the Thirty-First SIGCSE Technical Symposium on Computer Science Education,* Austin, TX, United States, 280-284. Retrieved September 4, 2006 from http://doi.acm.org/10.1145/330908.331870

Alphonce, C., & Ventura, P. (2002). Object orientation in CS1-CS2 by design. *ITiCSE '02: Proceedings of the 7th Annual Conference on Innovation and Technology in Computer Science Education,* Aarhus, Denmark. 70-74, Retrieved September 4, 2006 from http://doi.acm.org/10.1145/544414.544437

Aly, A. A., & Akhtar, S. (2004). Cryptography and security protocols course for undergraduate IT students. *SIGCSE Bulletin (Association for Computing Machinery, Special Interest Group on Computer Science Education), 36*(2), 44-47.

Anderson, R., Dickey, M., & Perkins, H. (2001). Experiences with tutored video instruction for introductory programming courses. *SIGCSE '01: Proceedings of the Thirty-Second SIGCSE Technical Symposium on Computer Science Education,* Charlotte, NC, United States, 347-351. Retrieved September 4, 2006 from http://doi.acm.org/10.1145/364447.364619

Anttila, I., Jormanainen, I., Kannusmäki, O. and Lehtonen, J. (2001). Lego-Compiler. Kolin Kolistelut – Koli Calling 2001, Proceedings of the First Annual Finnish Annual Finnish Baltic Sea Conference on Computer Science Education, Koli, Finland, 9-12. Retrieved September 4, 2006 from http://cs.joensuu.fi/kolistelut/archive/2001/koli_proc_2001.pdf

Applin, A. G. (2001). Second language acquisition and CS1. *SIGCSE '01: Proceedings of the Thirty-Second SIGCSE Technical Symposium on Computer Science Education,* Charlotte, NC, United States, 174-178. Retrieved September 4, 2006 from http://doi.acm.org/10.1145/364447.364579

Ariga, T., & Tsuiki, H. (2001). Programming for students of information design. *SIGCSE Bulletin (Association for Computing Machinery, Special Interest Group on Computer Science Education), 33*(4), 59-63.

Armen, C., & Morelli, R. (2005). Teaching about the risks of electronic voting technology. *ITiCSE '05: Proceedings of the 10th Annual SIGCSE Conference on Innovation and Technology in Computer Science Education,* Caparica, Portugal, 227-231. Retrieved September 4, 2006 from http://doi.acm.org/10.1145/1067445.1067508

Aycock, J., & Uhl, J. (2005). Choice in the classroom. *SIGCSE Bulletin (Association for Computing Machinery, Special Interest Group on Computer Science Education), 37*(4), 84-88.

Bagert, D., & Mead, N. (2001). Software engineering as a professional discipline. *Computer Science Education, 11*(1), 73-87.

Bailey, T., & Forbes, J. (2005). Just-in-time teaching for CS0. *SIGCSE '05: Proceedings of the 36th SIGCSE Technical Symposium on Computer Science Education,* St. Louis, MO, United States, 366-370. Retrieved September 4, 2006 from http://doi.acm.org/10.1145/1047344.1047469

Baldwin, D. (2000). Some thoughts on undergraduate teaching and the ph.D. *SIGCSE Bulletin (Association for Computing Machinery, Special Interest Group on Computer Science Education), 32*(4), 60-62.

Barker, L., & Garvin-Doxas, K. (2004). Making Visible the Behaviors that Influence Learning Environment: A Qualitative Exploration of Computer Science Classrooms. *Computer Science Education, 14*(2), 119-145.

Barker, L. J., Garvin-Doxas, K., & Roberts, E. (2005). What can computer science learn from a fine arts approach to teaching? *SIGCSE '05: Proceedings of the 36th SIGCSE Technical Symposium on Computer Science Education,* St. Louis, MO, United States, 421-425. Retrieved September 4, 2006 from http://doi.acm.org/10.1145/1047344.1047482

Beck, L. L., Chizhik, A. W., & McElroy, A. C. (2005). Cooperative learning techniques in CS1: Design and experimental evaluation. *SIGCSE '05: Proceedings of the 36th SIGCSE Technical Symposium on Computer Science Education,* St. Louis, MO, United States, 470-474. Retrieved September 4, 2006 from http://doi.acm.org/10.1145/1047344.1047495

Becker, B. W. (2001). Teaching CS1 with Karel the robot in java. *SIGCSE '01: Proceedings of the Thirty-Second SIGCSE Technical Symposium on Computer Science Education,* Charlotte, NC, United States, 50-54. Retrieved September 4, 2006 from http://doi.acm.org/10.1145/364447.364536

Bednarik, R. & Fränti, P. (2004). Survival of students with different learning preferences. *Kolin Kolistelut – Koli Calling 2004, Proceedings of the Fourth Annual Finnish Annual Finnish Baltic Sea Conference on Computer Science Education,* Koli, Finland, 121-125. Retrieved September 4, 2006 from http://cs.joensuu.fi/kolistelut/archive/2004/koli_proc_2004.pdf

Bennedsen, J. & Caspersen, M. E. (2005). An upcoming study of potential success factors for an introductory model-driven programming course. *Kolin Kolistelut – Koli Calling 2005, Proceedings of the Fourth Annual Finnish Annual Finnish Baltic Sea Conference on Computer Science Education,* Koli, Finland, 166-169. Retrieved September 4, 2006 from http://cs.joensuu.fi/kolistelut/archive/2005/koli_proc_2005.pdf

Beyer, S., Rynes, K., Perrault, J., Hay, K., & Haller, S. (2003). Gender differences in computer science students. *SIGCSE '03: Proceedings of the 34th SIGCSE Technical Symposium on Computer Science Education,* Reno, NV, United States, 49-53. Retrieved September 4, 2006 from http://doi.acm.org/10.1145/611892.611930

Boehm, B., Port, D., & Winsor Brown, A. (2002). Balancing plan-driven and agile methods in software engineering project courses. *Computer Science Education, 12*(3), 187-195.

Booth, S. (2001). Learning computer science and engineering in context. *Computer Science Education, 11*(3), 169-188.

Bouvier, D. J. (2003). Pilot study: Living flowcharts in an introduction to programming course. *SIGCSE '03: Proceedings of the 34th SIGCSE Technical Symposium on Computer Science Education,* Reno, NV, United States, 293-295. Retrieved September 4, 2006 from http://doi.acm.org/10.1145/611892.611991

Bruce, K. B., Danyluk, A., & Murtagh, T. (2001). A library to support a graphics-based object-first approach to CS 1. *SIGCSE '01: Proceedings of the Thirty-Second SIGCSE Technical Symposium on Computer Science Education,* Charlotte, NC, United States, 6-10. Retrieved September 4, 2006 from http://doi.acm.org/10.1145/364447.364527

Bruhn, R. E., & Burton, P. J. (2003). An approach to teaching java using computers. *SIGCSE Bulletin (Association for Computing Machinery, Special Interest Group on Computer Science Education), 35*(4), 94-99.

Bruhn, R. E., & Camp, J. (2004). Capstone course creates useful business products and corporate-ready students. *SIGCSE Bulletin (Association for Computing Machinery, Special Interest Group on Computer Science Education), 36*(2), 87-92.

Bryant, K. C. (2004}). The evaluation of courses in information systems. *Sixth Australasian Computing Education Conference (ACE2004); Conferences in Research and Practice in Information Technology, 30,* 15-23.

Buck, D., & Stucki, D. J. (2001). JKarelRobot: A case study in supporting levels of cognitive development in the computer science curriculum. *SIGCSE '01: Proceedings of the Thirty-Second SIGCSE Technical Symposium on Computer Science Education,* Charlotte, NC, United States, 16-20. Retrieved September 4, 2006 from http://doi.acm.org/10.1145/364447.364529

Butler, J. E., & Brockman, J. B. (2001). A web-based learning tool that simulates a simple computer architecture. *SIGCSE Bulletin (Association for Computing Machinery, Special Interest Group on Computer Science Education), 33*(2), 47-50.

Carr, S., Chen, P., Jozwowski, T. R., Mayo, J., & Shene, C. (2002). Channels, visualization, and topology editor. *ITiCSE '02: Proceedings of the 7th Annual Conference on Innovation and Technology in Computer Science Education,* Aarhus, Denmark, 106-110. Retrieved September 4, 2006 from http://doi.acm.org/10.1145/544414.544448

Carr, S., & Shene, C. (2000). A portable class library for teaching multithreaded programming. *ITiCSE '00: Proceedings of the 5th Annual SIGCSE/SIGCUE ITiCSE conference on Innovation and Technology in Computer Science Education,* Helsinki, Finland, 124-127. Retrieved September 4, 2006 from http://doi.acm.org/10.1145/343048.343138

Cavalcante, R., Finley, T., & Rodger, S. H. (2004). A visual and interactive automata theory course with JFLAP 4.0. *SIGCSE '04: Proceedings of the 35th SIGCSE Technical Symposium on Computer Science Education,* Norfolk, VA, United States, 140-144. Retrieved September 4, 2006 from http://doi.acm.org/10.1145/971300.971349

Cernuda del Río, A. (2004). How not to go about a programming assignment. *SIGCSE Bulletin (Association for Computing Machinery, Special Interest Group on Computer Science Education), 36*(2), 97-100.

Chalk, P. (2000). Webworlds—web-based modeling environments for learning software engineering. *Computer Science Education, 10*(1), 39-56.

Chamillard, A. T., & Joiner, J. K. (2001). Using lab practica to evaluate programming ability. *SIGCSE '01: Proceedings of the Thirty-Second SIGCSE Technical Symposium on Computer Science Education,* Charlotte, NC, United States, 159-163. Retrieved September 4, 2006 from http://doi.acm.org/10.1145/364447.364572

Chamillard, A. T., & Merkle, L. D. (2002). Management challenges in a large introductory computer science course. *SIGCSE '02: Proceedings of the 33rd SIGCSE Technical Symposium on Computer Science Education,* Cincinnati, OH, United States 252-256. Retrieved September 4, 2006 from http://doi.acm.org/10.1145/563340.563440

Chamillard, A. T., Moore, J. A., & Gibson, D. S. (2002, February).Using graphics in an introductory computer science course. *Journal of Computer Science Education Online.* Retrieved September 4, 2006 from http://www.iste.org/Template.cfm?Section=February&Template=/MembersOnly.cfm&ContentID=4244

Chandra, S. (2003). Beacond: A peer-to-peer system to teach ubiquitous computing. *SIGCSE '03: Proceedings of the 34th SIGCSE Technical Symposium on Computer Science Education,* Reno, NV, United States, 257-261. Retrieved September 4, 2006 from http://doi.acm.org/10.1145/611892.611983

Chen, S., & Morris, S. (2005). Iconic programming for flowcharts, java, turing, etc. *ITiCSE '05: Proceedings of the 10th Annual SIGCSE Conference on Innovation and Technology in Computer Science Education,* Caparica, Portugal, 104-107. Retrieved September 4, 2006 from http://doi.acm.org/10.1145/1067445.1067477

Christensen, H. B., rbak, & Caspersen, M. E. (2002). Frameworks in CS1: A different way of introducing event-driven programming. *ITiCSE '02: Proceedings of the 7th Annual Conference on Innovation and Technology in Computer Science Education,* Aarhus, Denmark, 75-79. Retrieved September 4, 2006 from http://doi.acm.org/10.1145/544414.544438

Cigas, J. (2003). An introductory course in network administration. *SIGCSE '03: Proceedings of the 34th SIGCSE Technical Symposium on Computer Science Education,* Reno, NV, United States, 113-116. Retrieved September 4, 2006 from http://doi.acm.org/10.1145/611892.611946

Cinnéide, M. Ó. & Tynan, R. (2004). A problem-based approach to teaching design patterns. *ITiCSE-WGR '04: Working Group Reports from ITiCSE on Innovation and Technology in Computer Science Education,* Leeds, United Kingdom, 80-82. Retrieved September 4, 2006 from http://doi.acm.org/10.1145/1044550.1041663

Clancy, M., Titterton, N., Ryan, C., Slotta, J., & Linn, M. (2003). New roles for students, instructors, and computers in a lab-based introductory programming course. *SIGCSE '03: Proceedings of the 34th SIGCSE Technical Symposium on Computer Science Education,* Reno, NV, United States, 132-136. Retrieved September 4, 2006 from http://doi.acm.org/10.1145/611892.611951

Clark, M. (2000). Getting participation through discussion. *SIGCSE '00: Proceedings of the Thirty-First SIGCSE Technical Symposium on Computer Science Education,* Austin, TX, United States, 129-133. Retrieved September 4, 2006 from http://doi.acm.org/10.1145/330908.331841

Clark, N. (2005}). Evaluating student teams developing unique industry projects. *Seventh Australasian Computing Education Conference (ACE2005); Conferences in Research and Practice in Information Technology, 42,* 21-30.

Clark, N. (2004}). Peer testing in software engineering projects. *Sixth Australasian Computing Education Conference (ACE2004); Conferences in Research and Practice in Information Technology, 30,* 41-48.

Claypool, K., & Claypool, M. (2005). Teaching software engineering through game design. *ITiCSE '05: Proceedings of the 10th Annual SIGCSE Conference on Innovation and Technology in Computer Science Education,* Caparica, Portugal, 123-127. Retrieved September 4, 2006 from http://doi.acm.org/10.1145/1067445.1067482

Claypool, M., Finkel, D., & Wills, C. (2001). An open source laboratory for operating systems projects. *ITiCSE '01: Proceedings of the 6th Annual Conference on Innovation and Technology in Computer Science Education,* Canterbury, United Kingdom, 145-148. Retrieved September 4, 2006 from http://doi.acm.org/10.1145/377435.377669

Clayton, D., & Lynch, T. (2002). Ten years of strategies to increase participation of women in computing programs: The central queensland university experience: 1999--2001. *SIGCSE Bulletin (Association for Computing Machinery, Special Interest Group on Computer Science Education), 34*(2), 89-93.

Cohen, R. F., Fairley, A. V., Gerry, D., & Lima, G. R. (2005). Accessibility in introductory computer science. *SIGCSE '05: Proceedings of the 36th SIGCSE Technical Symposium on Computer Science Education,* St. Louis, MO, United States, 17-21. Retrieved September 4, 2006 from http://doi.acm.org/10.1145/1047344.1047367

Combs, W., Hawkins, R., Pore, T., Schechet, A., Wahls, T., & Ziantz, L. (2005). The course scheduling problem as a source of student projects. *SIGCSE '05: Proceedings of the 36th SIGCSE Technical Symposium on Computer Science Education,* St. Louis, MO, United States, 81-85. Retrieved September 4, 2006 from http://doi.acm.org/10.1145/1047344.1047385

Cooper, S., Dann, W., & Pausch, R. (2003). Using animated 3D graphics to prepare novices for CS1. *Computer Science Education, 13*(1), 3-30.

Coppit, D., & Haddox-Schatz, J. M. (2005). Large team projects in software engineering courses. *SIGCSE '05: Proceedings of the 36th SIGCSE Technical Symposium on Computer Science Education,* St. Louis, MO, United States, 137-141. Retrieved September 4, 2006 from http://doi.acm.org/10.1145/1047344.1047400

Countermine, T., & Pfeiffer, P. (2000). Implementing an IT concentration in a CS department: Content, rationale, and initial impact. *SIGCSE '00: Proceedings of the Thirty-First SIGCSE Technical Symposium on Computer Science Education,* Austin, TX, United States, 275-279. Retrieved September 4, 2006 from http://doi.acm.org/10.1145/330908.331869

Crouch, D. B., & Schwartzman, L. (2003). Computer science accreditation: The advantages of being different. *SIGCSE '03: Proceedings of the 34th SIGCSE Technical Symposium on Computer Science Education,* Reno, NV, United States, 36-40. Retrieved September 4, 2006 from http://doi.acm.org/10.1145/611892.611927

Cunningham, S. (2002). Graphical problem solving and visual communication in the beginning computer graphics course. *SIGCSE '02: Proceedings of the 33rd SIGCSE Technical Symposium on Computer Science Education,* Cincinnati, OH, United States 181-185. Retrieved September 4, 2006 from http://doi.acm.org/10.1145/563340.563410

Czajkowski, M. F., Foster, C. V., Hewett, T. T., Casacio, J. A., Regli, W. C., & Sperber, H. A. (2001). A student project in software evaluation. *ITiCSE '01: Proceedings of the 6th Annual Conference on Innovation and Technology in Computer Science Education,* Canterbury, United Kingdom, 13-16. Retrieved September 4, 2006 from http://doi.acm.org/10.1145/377435.377446

Daly, C., & Horgan, J. (2005). Patterns of plagiarism. *SIGCSE '05: Proceedings of the 36th SIGCSE Technical Symposium on Computer Science Education,* St. Louis, MO, United States, 383-387. Retrieved September 4, 2006 from http://doi.acm.org/10.1145/1047344.1047473

Dann, W., Dragon, T., Cooper, S., Dietzler, K., Ryan, K., & Pausch, R. (2003). Objects: Visualization of behavior and state. *ITiCSE '03: Proceedings of the 8th Annual Conference on Innovation and Technology in Computer Science Education,* Thessaloniki, Greece, 84-88. Retrieved September 4, 2006 from http://doi.acm.org/10.1145/961511.961537

Dantin, U. (2005}). Application of personas in user interface design for educational software. *Seventh Australasian Computing Education Conference (ACE2005); Conferences in Research and Practice in Information Technology, 4,2* 239-247.

D'Antonio, L. (2003). Incorporating bioinformatics in an algorithms course. *ITiCSE '03: Proceedings of the 8th Annual Conference on Innovation and Technology in Computer Science Education,* Thessaloniki, Greece, 211-214. Retrieved September 4, 2006 from http://doi.acm.org/10.1145/961511.961569

Davis, T., Geist, R., Matzko, S., & Westall, J. (2004). ?????: A first step. *SIGCSE '04: Proceedings of the 35th SIGCSE Technical Symposium on Computer Science Education,* Norfolk, VA, United States, 125-129. Retrieved September 4, 2006 from http://doi.acm.org/10.1145/971300.971345

Debray, S. (2004). Writing efficient programs: Performance issues in an undergraduate CS curriculum. *SIGCSE '04: Proceedings of the 35th SIGCSE Technical Symposium on Computer Science Education,* Norfolk, VA, United States, 275-279. Retrieved September 4, 2006 from http://doi.acm.org/10.1145/971300.971396

Debray, S. (2002). Making compiler design relevant for students who will (most likely) never design a compiler. *SIGCSE '02: Proceedings of the 33rd SIGCSE Technical Symposium on Computer Science Education,* Cincinnati, OH, United States 341-345. Retrieved September 4, 2006 from http://doi.acm.org/10.1145/563340.563473

Demetriadis, S., Triantfillou, E., & Pombortsis, A. (2003). A phenomenographic study of students' attitudes toward the use of multiple media for learning. *ITiCSE '03: Proceedings of the 8th Annual Conference on Innovation and Technology in Computer Science Education,* Thessaloniki, Greece, 183-187. Retrieved September 4, 2006 from http://doi.acm.org/10.1145/961511.961562

DePasquale, P., Lee, J. A. N., Pérez-Quiñones, M. A. (2004). Evaluation of subsetting programming language elements in a novice's programming environment. *SIGCSE '04: Proceedings of the 35th SIGCSE Technical Symposium on Computer Science Education,* Norfolk, VA, United States, 260-264. Retrieved September 4, 2006 from http://doi.acm.org/10.1145/971300.971392

deRaadt, M., Watson, R., & Toleman, M. (2003}). Language tug-of-war: Industry demand and academic choice. *Fifth Australasian Computing Education Conference (ACE2003); Conferences in Research and Practice in Information Technology, , 20* 137-142.

Dewan, P. (2005). Teaching inter-object design patterns to freshmen. *SIGCSE '05: Proceedings of the 36th SIGCSE Technical Symposium on Computer Science Education,* St. Louis, MO, United States, 482-486. Retrieved September 4, 2006 from http://doi.acm.org/10.1145/1047344.1047498

Dick, M. (2005). Student interviews as a tool for assessment and learning in a systems analysis and design course. *ITiCSE '05: Proceedings of the 10th Annual SIGCSE Conference on Innovation and Technology in Computer Science Education,* Caparica, Portugal, 24-28. Retrieved September 4, 2006 from http://doi.acm.org/10.1145/1067445.1067456

Dick, M., Sheard, J., & Markham, S. (2001). Is it okay to cheat? - the views of postgraduate students. *ITiCSE '01: Proceedings of the 6th Annual Conference on Innovation and Technology in Computer Science Education,* Canterbury, United Kingdom, 61-64. Retrieved September 4, 2006 from http://doi.acm.org/10.1145/377435.377474

Dierbach, C., Taylor, B., Zhou, H., & Zimand, I. (2005). Experiences with a CS0 course targeted for CS1 success. *SIGCSE '05: Proceedings of the 36th SIGCSE Technical Symposium on Computer Science Education,* St. Louis, MO, United States, 317-320. Retrieved September 4, 2006 from http://doi.acm.org/10.1145/1047344.1047453

Donaldson, J. L. (2001). Architecture-dependent operating system project sequence. *SIGCSE '01: Proceedings of the Thirty-Second SIGCSE Technical Symposium on Computer Science Education,* Charlotte, NC, United States, 322-326. Retrieved September 4, 2006 from http://doi.acm.org/10.1145/364447.364613

Drummond, S., Boldyreff, C., & Ramage, M. (2001). Evaluating groupware support for software engineering students. *Computer Science Education, 11*(1), 33-54.

Drury, H., Kay, J., & Losberg, W. (2003}). Student satisfaction with groupwork in undergraduate computer science : Do things get better? *Fifth Australasian Computing Education Conference (ACE2003); Conferences in Research and Practice in Information Technology, 20,* 77-85.

Edwards, S. H. (2004). Using software testing to move students from trial-and-error to reflection-in-action. *SIGCSE '04: Proceedings of the 35th SIGCSE Technical Symposium on Computer Science Education,* Norfolk, VA, United States, 26-30. from Retrieved September 4, 2006 http://doi.acm.org/10.1145/971300.971312

Efopoulos, V., Dagdilelis, V., Evangelidis, G., & Satratzemi, M. (2005). WIPE: A programming environment for novices. *ITiCSE '05: Proceedings of the 10th Annual SIGCSE Conference on Innovation and Technology in Computer Science Education,* Caparica, Portugal, 113-117. Retrieved September 4, 2006 from http://doi.acm.org/10.1145/1067445.1067479

Egea, K. (2003}). Managing the managers: Collaborative virtual teams with large staff and student numbers. *Fifth Australasian Computing Education Conference (ACE2003); Conferences in Research and Practice in Information Technology, 20,* 87-94.

Eisenberg, M. (2003). Creating a computer science canon: A course of "classic" readings in computer science. *SIGCSE '03: Proceedings of the 34th SIGCSE Technical Symposium on Computer*

Science Education, Reno, NV, United States, 336-340. Retrieved September 4, 2006 from http://doi.acm.org/10.1145/611892.612002

Elsharnouby, T., & Shankar, A. U. (2005). Using SeSFJava in teaching introductory network courses. *SIGCSE '05: Proceedings of the 36th SIGCSE Technical Symposium on Computer Science Education,* St. Louis, MO, United States, 67-71. Retrieved September 4, 2006 from http://doi.acm.org/10.1145/1047344.1047381

Ericson, B., Guzdial, M., & Biggers, M. (2005). A model for improving secondary CS education. *SIGCSE '05: Proceedings of the 36th SIGCSE Technical Symposium on Computer Science Education,* St. Louis, MO, United States, 332-336. Retrieved September 4, 2006 from http://doi.acm.org/10.1145/1047344.1047460

Faux, R. (2003, April). The role of observation in computer science learning. *Journal of Computer Science Education Online.* Retrieved September 4, 2006 from http://www.iste.org/Template.cfm?Section=April&Template=/MembersOnly.cfm&ContentID=4223

Fekete, A. (2003). Using counter-examples in the data structures course. *Fifth Australasian Computing Education Conference (ACE2003); Conferences in Research and Practice in Information Technology, 20,* 179-186.

Felleisen, M., Findler, R., Flatt, M., & Krishnamurthi, S. (2004). The TeachScheme! Project: Computing and programming for every student. *Computer Science Education, 14*(1), 55-77.

Fenwick, J. B., Norris, C., & Wilkes, J. (2002). Scientific experimentation via the matching game. *SIGCSE '02: Proceedings of the 33rd SIGCSE Technical Symposium on Computer Science Education,* Cincinnati, OH, United States 326-330. Retrieved September 4, 2006 from http://doi.acm.org/10.1145/563340.563469

Fernandes, E., & Kumar, A. N. (2004). A tutor on scope for the programming languages course. *SIGCSE '04: Proceedings of the 35th SIGCSE Technical Symposium on Computer Science Education,* Norfolk, VA, United States, 90-93. Retrieved September 4, 2006 from http://doi.acm.org/10.1145/971300.971332

Filho, W. P. P. (2001). Requirements for an educational software development process. *ITiCSE '01: Proceedings of the 6th Annual Conference on Innovation and Technology in Computer Science Education,* Canterbury, United Kingdom, 65-68. Retrieved September 4, 2006 from http://doi.acm.org/10.1145/377435.377476

Fincher, S., Petre, M., Tenenberg, J., Blaha, K., Bouvier, D., Chen, T-Z., et al. (2004). A multi-national, multi-institutional study of student-generated software designs. *Kolin Kolistelut – Koli Calling 2004, Proceedings of the Fourth Annual Finnish Annual Finnish Baltic Sea Conference on*

Computer Science Education, Koli, Finland, 11-19. Retrieved September 4, 2006 from

http://cs.joensuu.fi/kolistelut/archive/2004/koli_proc_2004.pdf

Fisher, J., Lowther, J., & Shene, C. (2004). If you know b-splines well, you also know NURBS! *SIGCSE '04: Proceedings of the 35th SIGCSE Technical Symposium on Computer Science Education,* Norfolk, VA, United States, 343-347. Retrieved September 4, 2006 from http://doi.acm.org/10.1145/971300.971420

Fleury, A. E. (2001). Encapsualtion and reuse as viewed by java students. *SIGCSE '01: Proceedings of the Thirty-Second SIGCSE Technical Symposium on Computer Science Education,* Charlotte, NC, United States, 189-193. Retrieved September 4, 2006 from http://doi.acm.org/10.1145/364447.364582

Francioni, J. M., & Smith, A. C. (2002). Computer science accessibility for students with visual disabilities. *SIGCSE '02: Proceedings of the 33rd SIGCSE Technical Symposium on Computer Science Education,* Cincinnati, OH, United States 91-95. Retrieved September 4, 2006 from http://doi.acm.org/10.1145/563340.563372

Frens, J. D. (2004). Taming the tiger: Teaching the next version of java. *SIGCSE '04: Proceedings of the 35th SIGCSE Technical Symposium on Computer Science Education,* Norfolk, VA, United States, 151-155. Retrieved September 4, 2006 from http://doi.acm.org/10.1145/971300.971356

Frieze, C., & Blum, L. (2002). Building an effective computer science student organization: The Carnegie Mellon Women@SCS action plan. *SIGCSE Bulletin (Association for Computing Machinery, Special Interest Group on Computer Science Education), 34*(2), 74-78.

Gal-Ezer, J., Vilner, T., & Zur, E. (2004). Teaching algorithm efficiency at CS1 level: A different approach. *Computer Science Education, 14*(3), 235-248.

Gal-Ezer, J., & Zeldes, A. (2000). Teaching software designing skills. *Computer Science Education, 10*(1), 25-38.

Gallagher, J. C., & Perretta, S. (2002). WWW autonomous robotics: Enabling wide area access to a computer engineering practicum. *SIGCSE '02: Proceedings of the 33rd SIGCSE Technical Symposium on Computer Science Education,* Cincinnati, OH, United States 13-17. Retrieved September 4, 2006 from http://doi.acm.org/10.1145/563340.563346

Galpin, V. (2002). Women in computing around the world. *SIGCSE Bulletin (Association for Computing Machinery, Special Interest Group on Computer Science Education), 34*(2), 94-100.

Gaona, A. L., (2000). The relevance of design in CS1. *SIGCSE Bulletin (Association for Computing Machinery, Special Interest Group on Computer Science Education), 32*(2), 53-55.

Garner, S., Haden, P., & Robins, A. (2005}). My program is correct but it doesn't run: A preliminary investigation of novice programmers' problems. *Seventh Australasian Computing*

Education Conference (ACE2005); Conferences in Research and Practice in Information Technology, 42, 173-180.

Gegg-Harrison, T. S. (2001). Ancient egyptian numbers: A CS-complete example. *SIGCSE '01: Proceedings of the Thirty-Second SIGCSE Technical Symposium on Computer Science Education,* Charlotte, NC, United States, 268-272. Retrieved September 4, 2006 from http://doi.acm.org/10.1145/364447.364598

George, C. E. (2002). Using visualization to aid program construction tasks. *SIGCSE '02: Proceedings of the 33rd SIGCSE Technical Symposium on Computer Science Education,* Cincinnati, OH, United States 191-195. Retrieved September 4, 2006 from http://doi.acm.org/10.1145/563340.563413

Gerdt, P. (2001). Applying computer supported collaborative writing in education. Kolin Kolistelut – Koli Calling 2001, Proceedings of the First Annual Finnish Annual Finnish Baltic Sea Conference on Computer Science Education, Koli, Finland, 38-45. Retrieved September 4, 2006 from http://cs.joensuu.fi/kolistelut/archive/2001/koli_proc_2001.pdf

Gerhardt-Powals, J., & Powals, M. H. (2000). Distance education: Law attempts to catch up with technology (battle between copyright owners and academics). *SIGCSE '00: Proceedings of the Thirty-First SIGCSE Technical Symposium on Computer Science Education,* Austin, TX, United States, 335-342. Retrieved September 4, 2006 from http://doi.acm.org/10.1145/330908.331881

Gibson, J. P., & O'Kelly, J. (2005). Software engineering as a model of understanding for learning and problem solving. *ICER '05: Proceedings of the 2005 International Workshop on Computing Education Research,* Seattle, WA, United States, 87-97. Retrieved September 4, 2006 from http://doi.acm.org/10.1145/1089786.1089795

Ginat, D. (2004). Embedding instructive assertions in program design. *ITiCSE '04: Proceedings of the 9th Annual SIGCSE Conference on Innovation and Technology in Computer Science Education,* Leeds, United Kingdom, 62-66. Retrieved September 4, 2006 from http://doi.acm.org/10.1145/1007996.1008015

Ginat, D. (2003). The greedy trap and learning from mistakes. *SIGCSE '03: Proceedings of the 34th SIGCSE Technical Symposium on Computer Science Education,* Reno, NV, United States, 11-15. Retrieved September 4, 2006 from http://doi.acm.org/10.1145/611892.611920

Ginat, D. (2002). On varying perspectives of problem decomposition. *SIGCSE '02: Proceedings of the 33rd SIGCSE Technical Symposium on Computer Science Education,* Cincinnati, OH, United States 331-335. Retrieved September 4, 2006 from http://doi.acm.org/10.1145/563340.563470

Goldweber, M., Davoli, R., & Morsiani, M. (2005). The Kaya OS project and the *μ*MPS hardware emulator. *ITiCSE '05: Proceedings of the 10th Annual SIGCSE Conference on Innovation and Technology in Computer Science Education,* Caparica, Portugal, 49-53. Retrieved September 4, 2006 from http://doi.acm.org/10.1145/1067445.1067462

Golub, E. (2004). Handwritten slides on a tabletPC in a discrete mathematics course. *SIGCSE '04: Proceedings of the 35th SIGCSE Technical Symposium on Computer Science Education,* Norfolk, VA, United States, 51-55. Retrieved September 4, 2006 from http://doi.acm.org/10.1145/971300.971322

Goold, A., & Coldwell, J. (2005). Teaching ethics in a virtual classroom. *ITiCSE '05: Proceedings of the 10th Annual SIGCSE Conference on Innovation and Technology in Computer Science Education,* Caparica, Portugal, 232-236. Retrieved September 4, 2006 from http://doi.acm.org/10.1145/1067445.1067509

Gotterbarn, D., & Clear, T. (2004}). Using SoDIS as a risk analysis process: A teaching perspective. *Sixth Australasian Computing Education Conference (ACE2004); Conferences in Research and Practice in Information Technology, 30,* 83-90.

Grandell, L., Peltomäki, M., & Salakoski, T. (2005). High school programming – A beyond-syntax analysis of novice programmers' difficulties. *Kolin Kolistelut – Koli Calling 2005, Proceedings of the Fourth Annual Finnish Annual Finnish Baltic Sea Conference on Computer Science Education,* Koli, Finland, 17-24. Retrieved September 4, 2006 from http://cs.joensuu.fi/kolistelut/archive/2005/koli_proc_2005.pdf

Grinder, M. T., Kim, S. B., Lutey, T. L., Ross, R. J., & Walsh, K. F. (2002). Loving to learn theory: Active learning modules for the theory of computing. *SIGCSE '02: Proceedings of the 33rd SIGCSE Technical Symposium on Computer Science Education,* Cincinnati, OH, United States 371-375. Retrieved September 4, 2006 from http://doi.acm.org/10.1145/563340.563488

Gruba, P., & Al-Mahmood, R. (2004}). Strategies for communication skills development. *Sixth Australasian Computing Education Conference (ACE2004); Conferences in Research and Practice in Information Technology, 30,* 101-107.

Gruba, P., Moffat, A., Sondergaard, H., & Zobel, J. (2004}). What drives curriculum change? *Sixth Australasian Computing Education Conference (ACE2004); Conferences in Research and Practice in Information Technology, , 30* 109-117.

Gupta, G. K. (2001). Information technology and liberal arts. *SIGCSE Bulletin (Association for Computing Machinery, Special Interest Group on Computer Science Education), 33*(2), 55-57.

Guzdial, M. (2001). Use of collaborative multimedia in computer science classes. *ITiCSE '01: Proceedings of the 6th Annual Conference on Innovation and Technology in Computer Science Education,*

Canterbury, United Kingdom, 17-20. Retrieved September 4, 2006 from
http://doi.acm.org/10.1145/377435.377452

Haberman, B., & Averbuch, H. (2002). The case of base cases: Why are they so difficult to
recognize? student difficulties with recursion. *ITiCSE '02: Proceedings of the 7th Annual Conference
on Innovation and Technology in Computer Science Education,* Aarhus, Denmark, 84-88. Retrieved
September 4, 2006 from http://doi.acm.org/10.1145/544414.544441

Hadjerrouit, S. (2001). Web-based application development: A software engineering approach.
*SIGCSE Bulletin (Association for Computing Machinery, Special Interest Group on Computer Science
Education), 33*(2), 31-34.

Hamelin, D. (2004). Searching the web to develop inquiry and collaborative skills. *ITiCSE-WGR
'04: Working Group Reports from ITiCSE on Innovation and Technology in Computer Science Education,*
Leeds, United Kingdom, 76-79. Retrieved September 4, 2006 from
http://doi.acm.org/10.1145/1044550.1041662

Hamer, J. (2004). An approach to teaching design patterns using musical composition. *ITiCSE '04:
Proceedings of the 9th Annual SIGCSE Conference on Innovation and Technology in Computer Science
Education,* Leeds, United Kingdom, 156-160. Retrieved September 4, 2006 from
http://doi.acm.org/10.1145/1007996.1008038

Hamey, L. G. C. (2003}). Teaching secure data communications using a game representation. *Fifth
Australasian Computing Education Conference (ACE2003); Conferences in Research and Practice in
Information Technology, 20,* 187-196.

Hanks, B., McDowell, C., Draper, D., & Krnjajic, M. (2004). Program quality with pair
programming in CS1. *ITiCSE '04: Proceedings of the 9th Annual SIGCSE Conference on Innovation
and Technology in Computer Science Education,* Leeds, United Kingdom. 176-180. Retrieved
September 4, 2006 from http://doi.acm.org/10.1145/1007996.1008043

Hannon, C., Huber, M., & Burnell, L. (2005). Research to classroom: Experiences from a
multi-institutional course in smart home technologies. *SIGCSE '05: Proceedings of the 36th
SIGCSE Technical Symposium on Computer Science Education,* St. Louis, MO, United States,
121-125. Retrieved September 4, 2006 from http://doi.acm.org/10.1145/1047344.1047395

Hansen, K. M., & Ratzer, A. V. (2002). Tool support for collaborative teaching and learning of
object-oriented modeling. *ITiCSE '02: Proceedings of the 7th Annual Conference on Innovation and
Technology in Computer Science Education,* Aarhus, Denmark,146-150. Retrieved September 4,
2006 from http://doi.acm.org/10.1145/544414.544458

Harrison, S. M. (2005). Opening the eyes of those who can see to the world of those who can't:
A case study. *SIGCSE '05: Proceedings of the 36th SIGCSE Technical Symposium on Computer Science*

Education, St. Louis, MO, United States, 22-26. Retrieved September 4, 2006 from
http://doi.acm.org/10.1145/1047344.1047368

Hartley, S. J. (2000). \"Alfonse, you have a message!\". *SIGCSE '00: Proceedings of the Thirty-First SIGCSE Technical Symposium on Computer Science Education,* Austin, TX, United States, 60-64. Retrieved September 4, 2006 from http://doi.acm.org/10.1145/330908.331813

Hazzan, O. (2005). Professional development workshop for female software engineers. *SIGCSE Bulletin (Association for Computing Machinery, Special Interest Group on Computer Science Education), 37*(2), 75-79.

Hazzan, O. (2003). Application of computer science ideas to the presentation of mathematical theorems and proofs. *SIGCSE Bulletin (Association for Computing Machinery, Special Interest Group on Computer Science Education), 35*(2), 38-42.

Hazzan, O. (2001). On the presentation of computer science problems. *SIGCSE Bulletin (Association for Computing Machinery, Special Interest Group on Computer Science Education), 33*(4), 55-58.

Herrmann, N., Popyack, J. L., Char, B., Zoski, P., Cera, C. D., & Lass, R. N., et al. (2003). Redesigning introductory computer programming using multi-level online modules for a mixed audience. *SIGCSE '03: Proceedings of the 34th SIGCSE Technical Symposium on Computer Science Education,* Reno, NV, United States, 196-200. Retrieved September 4, 2006 from http://doi.acm.org/10.1145/611892.611967

Hilburn, T. B., & Townhidnejad, M. (2000). Software quality: A curriculum postscript? *SIGCSE '00: Proceedings of the Thirty-First SIGCSE Technical Symposium on Computer Science Education,* Austin, TX, United States. 167-171, Retrieved September 4, 2006 from http://doi.acm.org/10.1145/330908.331848

Hislop, G., Lutz, M., Fernando Naveda, J., Michael McCracken, W., Mead, N., & Williams, L. (2002). Integrating agile practices into software engineering courses. *Computer Science Education, 12*(3), 169-185.

Hoffman, M. E., & Vance, D. R. (2005). Computer literacy: What students know and from whom they learned it. *SIGCSE '05: Proceedings of the 36th SIGCSE Technical Symposium on Computer Science Education,* St. Louis, MO, United States, 356-360. Retrieved September 4, 2006 from http://doi.acm.org/10.1145/1047344.1047467

Hogan, J. M., Smith, G., & Thomas, R. C. (2005}). Tight spirals and industry clients: The modern SE education experience. *Seventh Australasian Computing Education Conference (ACE2005); Conferences in Research and Practice in Information Technology, 42,* 217-222.

Holland, D. A., Lim, A. T., & Seltzer, M. I. (2002). A new instructional operating system. *SIGCSE '02: Proceedings of the 33rd SIGCSE Technical Symposium on Computer Science Education,* Cincinnati,

OH, United States 111-115. Retrieved September 4, 2006 from
http://doi.acm.org/10.1145/563340.563383

Holmboe, C. (2005). The linguistics of object-oriented design: Implications for teaching. *ITiCSE '05: Proceedings of the 10th Annual SIGCSE Conference on Innovation and Technology in Computer Science Education,* Caparica, Portugal, 188-192. Retrieved September 4, 2006 from http://doi.acm.org/10.1145/1067445.1067498

Hood, C. S., & Hood, D. J. (2005). Teaching programming and language concepts using LEGOs\&\#174; *ITiCSE '05: Proceedings of the 10th Annual SIGCSE Conference on Innovation and Technology in Computer Science Education,* Caparica, Portugal, 19-23. Retrieved September 4, 2006 from http://doi.acm.org/10.1145/1067445.1067454

Howe, E., Thornton, M., & Weide, B. W. (2004). Components-first approaches to CS1/CS2: Principles and practice. *SIGCSE '04: Proceedings of the 35th SIGCSE Technical Symposium on Computer Science Education,* Norfolk, VA, United States, 291-295. Retrieved September 4, 2006 from http://doi.acm.org/10.1145/971300.971404

Hsia, J. I., Simpson, E., Smith, D., & Cartwright, R. (2005). Taming java for the classroom. *SIGCSE '05: Proceedings of the 36th SIGCSE Technical Symposium on Computer Science Education,* St. Louis, MO, United States, 327-331. Retrieved September 4, 2006 from http://doi.acm.org/10.1145/1047344.1047459

Hu, C. (2003). A framework for applet animations with controls. *SIGCSE Bulletin (Association for Computing Machinery, Special Interest Group on Computer Science Education), 35*(4), 90-93.

Hunt, K. (2003). Using image processing to teach CS1 and CS2. *SIGCSE Bulletin (Association for Computing Machinery, Special Interest Group on Computer Science Education), 35*(4), 86-89.

Huth, M. (2004). Mathematics for the exploration of requirements. *SIGCSE Bulletin (Association for Computing Machinery, Special Interest Group on Computer Science Education), 36*(2), 34-39.

Hämäläien, W. (2003). Problem-based learning of theoretical computer science. *Kolin Kolistelut – Koli Calling 2003, Proceedings of the Third Annual Finnish Annual Finnish Baltic Sea Conference on Computer Science Education,* Koli, Finland, 50-58. Retrieved September 4, 2006 from http://cs.joensuu.fi/kolistelut/archive/2003/koli_proc_2003.pdf

Hämäläinen, W., Myller, N., Pitkänen, S. H., & Lopéz-Cuadrado, J. (2002). Learning by gaming in apaptive learning system. *Kolin Kolistelut – Koli Calling 2002, Proceedings of the Second Annual Finnish Annual Finnish Baltic Sea Conference on Computer Science Education,* Koli, Finland, 22-26. Retrieved September 4, 2006 from http://cs.joensuu.fi/kolistelut/archive/2002/koli_proc_2002.pdf

Jacobsen, C. L., & Jadud, M. C. (2005). Towards concrete concurrency: Occam-pi on the LEGO mindstorms. *SIGCSE '05: Proceedings of the 36th SIGCSE Technical Symposium on Computer Science*

Education, St. Louis, MO, United States, 431-435. Retrieved September 4, 2006 from
http://doi.acm.org/10.1145/1047344.1047485

Jacobson, N. (2000). Using on-computer exams to ensure beginning students' programming
competency. *SIGCSE Bulletin (Association for Computing Machinery, Special Interest Group on
Computer Science Education), 32*(4), 53-56.

Jacobson, N., & Thornton, A. (2004). It is time to emphasize arraylists over arrays in java-based
first programming courses. *ITiCSE-WGR '04: Working Group Reports from ITiCSE on Innovation
and Technology in Computer Science Education,* Leeds, United Kingdom, 88-92. Retrieved
September 4, 2006 from http://doi.acm.org/10.1145/1044550.1041666

Jadud, M. (2005). A first look at novice compilation behaviour using BlueJ. *Computer Science
Education, 15*(1), 25-40.

Jalloul, G. (2000). Links: A framework for object-oriented software engineering. *Computer Science
Education, 10*(1), 75-93.

James, R. H. (2005). External sponsored projects: Lessons learned. *SIGCSE Bulletin (Association for
Computing Machinery, Special Interest Group on Computer Science Education), 37*(2), 94-98.

Jenkins, T. (2001). The motivation of students of programming. *ITiCSE '01: Proceedings of the 6th
Annual Conference on Innovation and Technology in Computer Science Education,* Canterbury, United
Kingdom, 53-56. Retrieved September 4, 2006 from
http://doi.acm.org/10.1145/377435.377472

Jepson, A., & Perl, T. (2002). Priming the pipeline. *SIGCSE Bulletin (Association for Computing
Machinery, Special Interest Group on Computer Science Education), 34*(2), 36-39.

Jiménez-Díaz, G., Gómez-Albarrán, M., et al. (2005). Software behaviour understanding supported
by dynamic visualization and role-play. *ITiCSE '05: Proceedings of the 10th Annual SIGCSE
Conference on Innovation and Technology in Computer Science Education,* Caparica, Portugal, 54-58.
Retrieved September 4, 2006 Retrieved September 4, 2006 from
http://doi.acm.org/10.1145/1067445.1067464

Johnson, D., & Caristi, J. (2002). Using extreme programming in the software design course.
Computer Science Education, 12(3), 223-234.

Jones, E. L. (2001). Integrating testing into the curriculum \— arsenic in small doses. *SIGCSE '01:
Proceedings of the Thirty-Second SIGCSE Technical Symposium on Computer Science Education,* Charlotte,
NC, United States, 337-341. Retrieved September 4, 2006 from
http://doi.acm.org/10.1145/364447.364617

Jones, E. L., & Allen, C. S. (2003). Repositories for CS courses: An evolutionary tale. *ITiCSE '03:
Proceedings of the 8th Annual Conference on Innovation and Technology in Computer Science Education,*

Thessaloniki, Greece, 119-123. Retrieved September 4, 2006 from
http://doi.acm.org/10.1145/961511.961546

Jones, M. (2005). The pedagogic opportunities of touch-screen voting. *ITiCSE '05: Proceedings of the 10th Annual SIGCSE Conference on Innovation and Technology in Computer Science Education,* Caparica, Portugal, 223-226. Retrieved September 4, 2006 from
http://doi.acm.org/10.1145/1067445.1067507

Jones, R. M. (2000). Design and implementation of computer games: A capstone course for undergraduate computer science education. *SIGCSE '00: Proceedings of the Thirty-First SIGCSE Technical Symposium on Computer Science Education,* Austin, TX, United States. 260-264, Retrieved September 4, 2006 from http://doi.acm.org/10.1145/330908.331866

Jormanainen, I. (2004). A visual approach for concretizing sorting algorithms. *Kolin Kolistelut – Koli Calling 2004, Proceedings of the Fourth Annual Finnish Annual Finnish Baltic Sea Conference on Computer Science Education,* Koli, Finland, 141-145. Retrieved September 4, 2006 from
http://cs.joensuu.fi/kolistelut/archive/2004/koli_proc_2004.pdf

Jovanovic-Dolecek, G., & Champac, V. H. (2000). CGTDEMO \— educational software for the central limit theorem. *SIGCSE Bulletin (Association for Computing Machinery, Special Interest Group on Computer Science Education), 32*(2), 46-48.

Joyce, D., & Young, A. L. (2004}). Developing and implementing a professional doctorate in computing. *Sixth Australasian Computing Education Conference (ACE2004); Conferences in Research and Practice in Information Technology, 30,* 145-149.

Kamppuri, M., Tedre, M. & Tukianen, M. (2005). A cultural approach to interface design. *Kolin Kolistelut – Koli Calling 2005, Proceedings of the Fourth Annual Finnish Annual Finnish Baltic Sea Conference on Computer Science Education,* Koli, Finland, 149-152. Retrieved September 4, 2006 from http://cs.joensuu.fi/kolistelut/archive/2005/koli_proc_2005.pdf

Karavita, V., Korhonen, A., Nikander, J., & Tenhunen, P. (2002). Effortless creation of algorithm visualization. *Kolin Kolistelut – Koli Calling 2002, Proceedings of the Second Annual Finnish Annual Finnish Baltic Sea Conference on Computer Science Education,* Koli, Finland, 52-56. Retrieved September 4, 2006 from
http://cs.joensuu.fi/kolistelut/archive/2002/koli_proc_2002.pdf

Keefe, K., & Dick, M. (2004}). Using extreme programming in a capstone project. *Sixth Australasian Computing Education Conference (ACE2004); Conferences in Research and Practice in Information Technology, 30,* 151-160.

Kellerer, W., Autenrieth, A., & Iselt, A. (2000). Experiences with evaluation of SDL-based protocol engineering in education. *Computer Science Education, 10*(3), 225-241.

Kenny, C., & Pahl, C. (2005). Automated tutoring for a database skills training environment. *SIGCSE '05: Proceedings of the 36th SIGCSE Technical Symposium on Computer Science Education,* St. Louis, MO, United States. 58-62, Retrieved September 4, 2006 from http://doi.acm.org/10.1145/1047344.1047377

Khuri, S., & Holzapfel, K. (2001). EVEGA: An educational visulalization environment for graph algorithms. *ITiCSE '01: Proceedings of the 6th Annual Conference on Innovation and Technology in Computer Science Education,* Canterbury, United Kingdom, 101-104. Retrieved September 4, 2006 from http://doi.acm.org/10.1145/377435.377497

Khuri, S., & Hsu, H. (2000). Interactive packages for learning image compression algorithms. *ITiCSE '00: Proceedings of the 5th Annual SIGCSE/SIGCUE ITiCSEconference on Innovation and Technology in Computer Science Education,* Helsinki, Finland, 73-76. Retrieved September 4, 2006 from http://doi.acm.org/10.1145/343048.343081

Klassner, F. (2004). Enhancing lisp instruction with RCXLisp and robotics. *SIGCSE '04: Proceedings of the 35th SIGCSE Technical Symposium on Computer Science Education,* Norfolk, VA, United States, 214-218. Retrieved September 4, 2006 from http://doi.acm.org/10.1145/971300.971377

Koldehofe, B., & Tsigas, P. (2001). Using actors in an interactive animation in a graduate course on distributed system. *ITiCSE '01: Proceedings of the 6th Annual Conference on Innovation and Technology in Computer Science Education,* Canterbury, United Kingdom, 149-152. Retrieved September 4, 2006 from http://doi.acm.org/10.1145/377435.377670

Kolikant, Y. B. (2005). Students' alternative standards for correctness. *ICER '05: Proceedings of the 2005 International Workshop on Computing Education Research,* Seattle, WA, United States, 37-43. Retrieved September 4, 2006 from http://doi.acm.org/10.1145/1089786.1089790

Kolikant, Y. B., Ben-Ari, M., & Pollack, S. (2000). The anthropology semaphores. *ITiCSE '00: Proceedings of the 5th Annual SIGCSE/SIGCUE ITiCSE conference on Innovation and Technology in Computer Science Education,* Helsinki, Finland, 21-24. Retrieved September 4, 2006 from http://doi.acm.org/10.1145/343048.343061

Kolikant, Y. B., & Pollack, S. (2004). Community-oriented pedagogy for in-service CS teacher training. *ITiCSE '04: Proceedings of the 9th Annual SIGCSE Conference on Innovation and Technology in Computer Science Education,* Leeds, United Kingdom, 191-195. Retrieved September 4, 2006 from http://doi.acm.org/10.1145/1007996.1008047

Kolikant, Y., & Pollack, S. (2004). Establishing computer science professional norms among high-school students. *Computer Science Education, 14*(1), 21-35.

Korhonen, A., & Malmi, L. (2000). Algorithm simulation with automatic assessment. *ITiCSE '00: Proceedings of the 5th Annual SIGCSE/SIGCUE ITiCSEconference on Innovation and Technology in*

Computer Science Education, Helsinki, Finland, 160-163. Retrieved September 4, 2006 from http://doi.acm.org/10.1145/343048.343157

Korhonen, A., Malmi, L., Myllyselk\&\#228, P., Scheinin, , & Patrik. (2002). Does it make a difference if students exercise on the web or in the classroom? *ITiCSE '02: Proceedings of the 7th Annual Conference on Innovation and Technology in Computer Science Education,* Aarhus, Denmark, 121-124. Retrieved September 4, 2006 from http://doi.acm.org/10.1145/544414.544452

Kuittinen, M., & Sajaniemi, J. (2004). Teaching roles of variables in elementary programming courses. *ITiCSE '04: Proceedings of the 9th Annual SIGCSE Conference on Innovation and Technology in Computer Science Education,* Leeds, United Kingdom, 57-61. Retrieved September 4, 2006 from http://doi.acm.org/10.1145/1007996.1008014

Kumar, A. (2000). Dynamically generating problems on static scope. *ITiCSE '00: Proceedings of the 5th Annual SIGCSE/SIGCUE ITiCSE conference on Innovation and Technology in Computer Science Education,* Helsinki, Finland, 9-12. Retrieved September 4, 2006 from http://doi.acm.org/10.1145/343048.343055

Laakso, M-J., Salakoski, T., Korhonen, A. & Malmi, L. (2004). Automatic assessment of exercises for algorithms and data structures — a case study with TRAKLA2. *Kolin Kolistelut – Koli Calling 2004, Proceedings of the Fourth Annual Finnish Annual Finnish Baltic Sea Conference on Computer Science Education,* Koli, Finland, 28-36. Retrieved September 4, 2006 from http://cs.joensuu.fi/kolistelut/archive/2004/koli_proc_2004.pdf

Lai, Y. (2005). Teaching computer applications to pre-school teachers through problem based learning approach. *SIGCSE Bulletin (Association for Computing Machinery, Special Interest Group on Computer Science Education), 37*(4), 89-92.

Laine, H. (2001). SQO-Trainer. *Kolin Kolistelut – Koli Calling 2001, Proceedings of the First Annual Finnish Annual Finnish Baltic Sea Conference on Computer Science Education,* Koli, Finland, 13-19. Retrieved September 4, 2006 from http://cs.joensuu.fi/kolistelut/archive/2001/koli_proc_2001.pdf

Lane, H. C., & VanLehn, K. (2003). Coached program planning: Dialogue-based support for novice program design. *SIGCSE '03: Proceedings of the 34th SIGCSE Technical Symposium on Computer Science Education,* Reno, NV, United States, 148-152. Retrieved September 4, 2006 from http://doi.acm.org/10.1145/611892.611955

Lane, H. C., & VanLehn, K. (2005). Intention-based scoring: An approach to measuring success at solving the composition problem. *SIGCSE '05: Proceedings of the 36th SIGCSE Technical Symposium on Computer Science Education,* St. Louis, MO, United States, 373-377. Retrieved September 4, 2006 from http://doi.acm.org/10.1145/1047344.1047471

Lane, H. C., & VanLehn, K. (2005). Teaching the tacit knowledge of programming to noviceswith natural language tutoring. *Computer Science Education, 15*(3), 183-201.

Lapidot, T., & Hazzan, O. (2005). Song debugging: Merging content and pedagogy in computer science education. *SIGCSE Bulletin (Association for Computing Machinery, Special Interest Group on Computer Science Education), 37*(4), 79-83.

Lapidot, T., & Hazzan, O. (2003). Methods of teaching a computer science course for prospective teachers. *SIGCSE Bulletin (Association for Computing Machinery, Special Interest Group on Computer Science Education), 35*(4), 29-34.

La Russa, G. (2001). Concretising tools for computer science education. *Kolin Kolistelut – Koli Calling 2001, Proceedings of the First Annual Finnish Annual Finnish Baltic Sea Conference on Computer Science Education,* Koli, Finland, 4-8. Retrieved September 4, 2006 from http://cs.joensuu.fi/kolistelut/archive/2001/koli_proc_2001.pdf

La Russa, G., Silander, P., & Rytkönen, A. (2004). Utilizing pedagogical models in web-based CSE education. *Kolin Kolistelut – Koli Calling 2004, Proceedings of the Fourth Annual Finnish Annual Finnish Baltic Sea Conference on Computer Science Education,* Koli, Finland, 107-111. Retrieved September 4, 2006 from http://cs.joensuu.fi/kolistelut/archive/2004/koli_proc_2004.pdf

Lass, R. N., Cera, C. D., Bomberger, N. T., Char, B., Popyack, J. L., & Herrmann, N., et al. (2003). Tools and techniques for large scale grading using web-based commercial off-the-shelf software. *ITiCSE '03: Proceedings of the 8th Annual Conference on Innovation and Technology in Computer Science Education,* Thessaloniki, Greece, 168-172. Retrieved September 4, 2006 from http://doi.acm.org/10.1145/961511.961558

Last, M. Z., Daniels, M., Almstrum, V. L., Erickson, C., & Klein, B. (2000). An international student/faculty collaboration: The runestone project. *ITiCSE '00: Proceedings of the 5th Annual SIGCSE/SIGCUE ITiCSE conference on Innovation and Technology in Computer Science Education,* Helsinki, Finland, 128-131. Retrieved September 4, 2006 from http://doi.acm.org/10.1145/343048.343140

Latu, S., & Young, A. L. (2004}). Teaching ICT to pacific island background students. *Sixth Australasian Computing Education Conference (ACE2004); Conferences in Research and Practice in Information Technology , 30,* 169-175.

Learmonth, R. (2001). Flexible delivery of information systems as a core MBA subject. *ITiCSE '01: Proceedings of the 6th Annual Conference on Innovation and Technology in Computer Science Education,* Canterbury, United Kingdom, 29-32. Retrieved September 4, 2006 from http://doi.acm.org/10.1145/377435.377459

Lee, J. A. N. (2002). Internationalization of the curriculum report of a project within computer science. *SIGCSE '02: Proceedings of the 33rd SIGCSE Technical Symposium on Computer Science Education,* Cincinnati, OH, United States 68-72. Retrieved September 4, 2006 from http://doi.acm.org/10.1145/563340.563366

Lee, J. A. N. (2001). Teaching and learning in the 21st century: The development of "future CS faculty". *SIGCSE Bulletin (Association for Computing Machinery, Special Interest Group on Computer Science Education), 33*(2), 25-30.

Leventhal, L., & Barnes, J. (2003). Two for one: Squeezing human-computer interaction and software engineering into a core computer science course. *Computer Science Education, 13*(3), 177-190.

Leventhal, L. M., Barnes, J., & Chao, J. (2004). Term project user interface specifications in a United Statesbility engineering course: Challenges and suggestions. *SIGCSE '04: Proceedings of the 35th SIGCSE Technical Symposium on Computer Science Education,* Norfolk, VA, United States, 41-45. Retrieved September 4, 2006 from http://doi.acm.org/10.1145/971300.971316

Levitin, A. (2005). Analyze that: Puzzles and analysis of algorithms. *SIGCSE '05: Proceedings of the 36th SIGCSE Technical Symposium on Computer Science Education,* St. Louis, MO, United States, 171-175. Retrieved September 4, 2006 from http://doi.acm.org/10.1145/1047344.1047409

Levy, D., & Lapidot, T. (2002). Shared terminology, private syntax: The case of recursive descriptions. *ITiCSE '02: Proceedings of the 7th Annual Conference on Innovation and Technology in Computer Science Education,* Aarhus, Denmark, 89-93. Retrieved September 4, 2006 from http://doi.acm.org/10.1145/544414.544442

Lewis, T. L., Rosson, M. B., Pérez-Quiñones, M. A. (2004). What do the experts say?: Teaching introductory design from an expert's perspective. *SIGCSE '04: Proceedings of the 35th SIGCSE Technical Symposium on Computer Science Education,* Norfolk, VA, United States, 296-300. Retrieved September 4, 2006 from http://doi.acm.org/10.1145/971300.971405

Lister, R. (2004}). Teaching java first: Experiments with a pigs-early pedagogy. *Sixth Australasian Computing Education Conference (ACE2004); Conferences in Research and Practice in Information Technology, , 30* 177-183.

Lister, R. (2001). Objectives and objective assessment in CS1. *SIGCSE '01: Proceedings of the Thirty-Second SIGCSE Technical Symposium on Computer Science Education,* Charlotte, NC, United States, 292-296. Retrieved September 4, 2006 from http://doi.acm.org/10.1145/364447.364605

Lowther, J., & Shene, C. (2003). Teaching B-splines is not difficult! *SIGCSE '03: Proceedings of the 34th SIGCSE Technical Symposium on Computer Science Education,* Reno, NV, United States, 381-385. Retrieved September 4, 2006 from http://doi.acm.org/10.1145/611892.612012

Lukins, S., Levicki, A., & Burg, J. (2002). A tutorial program for propositional logic with human/computer interactive learning. *SIGCSE '02: Proceedings of the 33rd SIGCSE Technical Symposium on Computer Science Education,* Cincinnati, OH, United States 381-385. Retrieved September 4, 2006 from http://doi.acm.org/10.1145/563340.563490

Lynch, S., & Rajendran, K. (2005). Multi-agent systems design for novices. *Computer Science Education, 15*(1), 41-57.

Maj, S. P., Veal, D., & Duley, R. (2001). A proposed new high level abstraction for computer technology. *SIGCSE '01: Proceedings of the Thirty-Second SIGCSE Technical Symposium on Computer Science Education,* Charlotte, NC, United States, 199-203. Retrieved September 4, 2006 from http://doi.acm.org/10.1145/364447.364584

Manolopoulos, Y. (2005). On the number of recursive calls of recursive functions. *SIGCSE Bulletin (Association for Computing Machinery, Special Interest Group on Computer Science Education), 37*(2), 61-64.

Marks, J., Freeman, W., & Leitner, H. (2001). Teaching applied computing without programming: A case-based introductory course for general education. *SIGCSE '01: Proceedings of the Thirty-Second SIGCSE Technical Symposium on Computer Science Education,* Charlotte, NC, United States, 80-84. Retrieved September 4, 2006 from http://doi.acm.org/10.1145/364447.364547

Marrero, W., & Settle, A. (2005). Testing first: Emphasizing testing in early programming courses. *ITiCSE '05: Proceedings of the 10th Annual SIGCSE Conference on Innovation and Technology in Computer Science Education,* Caparica, Portugal, 4-8. Retrieved September 4, 2006 from http://doi.acm.org/10.1145/1067445.1067451

Matos, V., & Grasser, R. (2000). RELAX \— the relational algebra pocket calculator project. *SIGCSE Bulletin (Association for Computing Machinery, Special Interest Group on Computer Science Education), 32*(4), 40-44.

Maurer, W. D. (2002). The comparative programming languages course: A new chain of development. *SIGCSE '02: Proceedings of the 33rd SIGCSE Technical Symposium on Computer Science Education,* Cincinnati, OH, United States 336-340. Retrieved September 4, 2006 from http://doi.acm.org/10.1145/563340.563472

McCormick, J. W. (2005). We've been working on the railroad: A laboratory for real-time embedded systems. *SIGCSE '05: Proceedings of the 36th SIGCSE Technical Symposium on Computer*

Science Education, St. Louis, MO, United States, 530-534. Retrieved September 4, 2006 from http://doi.acm.org/10.1145/1047344.1047510

McGuffee, J. W. (2000). Defining computer science. *SIGCSE Bulletin (Association for Computing Machinery, Special Interest Group on Computer Science Education), 32*(2), 74-76.

McKinney, D., & Denton, L. F. (2005). Affective assessment of team skills in agile CS1 labs: The good, the bad, and the ugly. *SIGCSE '05: Proceedings of the 36th SIGCSE Technical Symposium on Computer Science Education,* St. Louis, MO, United States, 465-469. Retrieved September 4, 2006 from http://doi.acm.org/10.1145/1047344.1047494

Medley, M. D. (2001). Using qualitative research software for CS education research. *ITiCSE '01: Proceedings of the 6th Annual Conference on Innovation and Technology in Computer Science Education,* Canterbury, United Kingdom, 141-144. Retrieved September 4, 2006 from http://doi.acm.org/10.1145/377435.377668

Mendes, A. J., Gomes, A., Esteves, M., Marcelino, M. J., & Bravo, C., et al. (2005). Using simulation and collaboration in CS1 and CS2. *ITiCSE '05: Proceedings of the 10th Annual SIGCSE Conference on Innovation and Technology in Computer Science Education,* Caparica, Portugal, 193-197. Retrieved September 4, 2006 from http://doi.acm.org/10.1145/1067445.1067499

Mendes, E., Mosley, N., & Counsell, S. (2001). The cognitive flexibility theory0: An approach for teaching hypermedia engineering. *ITiCSE '01: Proceedings of the 6th Annual Conference on Innovation and Technology in Computer Science Education,* Canterbury, United Kingdom, 21-24. Retrieved September 4, 2006 from http://doi.acm.org/10.1145/377435.377457

Michael, M. (2000). Fostering and assessing communication skills in the computer science context. *SIGCSE '00: Proceedings of the Thirty-First SIGCSE Technical Symposium on Computer Science Education,* Austin, TX, United States, 119-123. Retrieved September 4, 2006 from http://doi.acm.org/10.1145/330908.331834

Miller, C. (2003). Relating theory to actual results in computer science and human-computer iInteraction. *Computer Science Education, 13*(3), 227-240.

Modak, V. D., Langan, D. D., & Hain, T. F. (2005). A pattern-based development tool for mobile agents. *SIGCSE '05: Proceedings of the 36th SIGCSE Technical Symposium on Computer Science Education,* St. Louis, MO, United States, 72-75. Retrieved September 4, 2006 from http://doi.acm.org/10.1145/1047344.1047382

Moody, D. L., & Sindre, G. (2003). Incorporating quality assurance processes into requirements analysis education. *ITiCSE '03: Proceedings of the 8th Annual Conference on Innovation and Technology in Computer Science Education,* Thessaloniki, Greece, 74-78. Retrieved September 4, 2006 from http://doi.acm.org/10.1145/961511.961534

Moore, T. K. (2002). Bringing the enterprise into a database systems course. *SIGCSE '02: Proceedings of the 33rd SIGCSE Technical Symposium on Computer Science Education,* Cincinnati, OH, United States 262-265. Retrieved September 4, 2006 from http://doi.acm.org/10.1145/563340.563443

Moritz, S. H., & Blank, G. D. (2005). A design-first curriculum for teaching java in a CS1 course. *SIGCSE Bulletin (Association for Computing Machinery, Special Interest Group on Computer Science Education), 37*(2), 89-93.

Moritz, S. H., Wei, F., Parvez, S. M., & Blank, G. D. (2005). From objects-first to design-first with multimedia and intelligent tutoring. *ITiCSE '05: Proceedings of the 10th Annual SIGCSE Conference on Innovation and Technology in Computer Science Education,* Caparica, Portugal, 99-103. Retrieved September 4, 2006 from http://doi.acm.org/10.1145/1067445.1067475

Morris, J. (2005}). Algorithm animation: Using the algorithm code to drive the animation. *Seventh Australasian Computing Education Conference (ACE2005); Conferences in Research and Practice in Information Technology, 42,* 15-20.

Mosiman, S., & Hiemcke, C. (2000). Interdisciplinary capstone group project: Designing autonomous race vehicles. *SIGCSE '00: Proceedings of the Thirty-First SIGCSE Technical Symposium on Computer Science Education,* Austin, TX, United States, 270-274. Retrieved September 4, 2006 from http://doi.acm.org/10.1145/330908.331868

Muller, O. (2005). Pattern oriented instruction and the enhancement of analogical reasoning. *ICER '05: Proceedings of the 2005 International Workshop on Computing Education Research,* Seattle, WA, United States, 57-67. Retrieved September 4, 2006 from http://doi.acm.org/10.1145/1089786.1089792

Muller, O., Haberman, B., & Averbuch, H. (2004). (An almost) pedagogical pattern for pattern-based problem-solving instruction. *ITiCSE '04: Proceedings of the 9th Annual SIGCSE Conference on Innovation and Technology in Computer Science Education,* Leeds, United Kingdom. 102-106. Retrieved September 4, 2006 from http://doi.acm.org/10.1145/1007996.1008025

Mäntylä, T. Laakso, M., Innola, E., & Salakoski, T. (2005). Student counseling with the SOPS-TOOl. *Kolin Kolistelut – Koli Calling 2005, Proceedings of the Fourth Annual Finnish Annual Finnish Baltic Sea Conference on Computer Science Education,* Koli, Finland, 119-124. Retrieved September 4, 2006 from http://cs.joensuu.fi/kolistelut/archive/2005/koli_proc_2005.pdf

Nagappan, N., Williams, L., Ferzli, M., Wiebe, E., Yang, K., & Miller, C., et al. (2003). Improving the CS1 experience with pair programming. *SIGCSE '03: Proceedings of the 34th SIGCSE Technical Symposium on Computer Science Education,* Reno, NV, United States, 359-362. Retrieved September 4, 2006 from http://doi.acm.org/10.1145/611892.612006

Nair, A. S. S., & Mahalakshmi, T. (2004). Conceptualizing data structures: A pedagogic approach. *ITiCSE-WGR '04: Working Group Reports from ITiCSE on Innovation and Technology in Computer Science Education,* Leeds, United Kingdom, 97-100. Retrieved September 4, 2006 from http://doi.acm.org/10.1145/1044550.1041668

Naumanen, M. (2003). A narrative perspective to students' experiences in problem-based leaning in theory of computation. *Kolin Kolistelut – Koli Calling 2003, Proceedings of the Third Annual Finnish Annual Finnish Baltic Sea Conference on Computer Science Education,* Koli, Finland, 59-62. Retrieved September 4, 2006 from http://cs.joensuu.fi/kolistelut/archive/2003/koli_proc_2003.pdf

Nerur, S., Ramanujan, S. & Kesh, S. (2002, April). Pedagogical issues in object orientation. *Journal of Computer Science Education Online.* Retrieved September 4, 2006 from http://www.iste.org/Template.cfm?Section=April1&Template=/MembersOnly.cfm&ContentID=4234

Nguyen, D., & Wong, S. B. (2001). Design patterns for sorting. *SIGCSE '01: Proceedings of the Thirty-Second SIGCSE Technical Symposium on Computer Science Education,* Charlotte, NC, United States, 263-267. Retrieved September 4, 2006 from http://doi.acm.org/10.1145/364447.364597

Noonan, R. E. (2000). An object-oriented view of backtracking. *SIGCSE '00: Proceedings of the Thirty-First SIGCSE Technical Symposium on Computer Science Education,* Austin, TX, United States, 362-366. Retrieved September 4, 2006 from http://doi.acm.org/10.1145/330908.331886

Noyes, J. L. (2002). A first course in computational science: (why a math book isn't enough). *SIGCSE '02: Proceedings of the 33rd SIGCSE Technical Symposium on Computer Science Education,* Cincinnati, OH, United States 18-22. Retrieved September 4, 2006 from http://doi.acm.org/10.1145/563340.563348

Null, L., & Rao, K. (2005). CAMERA: Introducing memory concepts via visualization. *SIGCSE '05: Proceedings of the 36th SIGCSE Technical Symposium on Computer Science Education,* St. Louis, MO, United States, 96-100. Retrieved September 4, 2006 from http://doi.acm.org/10.1145/1047344.1047389

Nørmark, K. (2000). A suite of WWW-based tools for advanced course management. *ITiCSE '00: Proceedings of the 5th Annual SIGCSE/SIGCUE ITiCSEconference on Innovation and Technology in Computer Science Education,* Helsinki, Finland, 65-68. Retrieved September 4, 2006 from http://doi.acm.org/10.1145/343048.343078

Odekirk, E., Jones, D., & Jensen, P. (2000). Three semesters of CSO using java: Assignments and experiences. *ITiCSE '00: Proceedings of the 5th Annual SIGCSE/SIGCUE ITiCSE conference on*

Innovation and Technology in Computer Science Education, Helsinki, Finland, 144-147. Retrieved September 4, 2006 from http://doi.acm.org/10.1145/343048.343148

Or-Bach, R., & Lavy, I. (2004). Cognitive activities of abstraction in object orientation: An empirical study. *SIGCSE Bulletin (Association for Computing Machinery, Special Interest Group on Computer Science Education), 36*(2), 82-86.

Ou, K.-L., Chen, G., Liu, C., & Liu, B. (2000). Instructional instruments for web group learning systems: The grouping, intervention, and strategy. *ITiCSE '00: Proceedings of the 5th Annual SIGCSE/SIGCUE ITiCSE conference on Innovation and Technology in Computer Science Education,* Helsinki, Finland, 69-72. Retrieved September 4, 2006 from http://doi.acm.org/10.1145/343048.343079

Parberry, I., Roden, T., & Kazemzadeh, M. B. (2005). Experience with an industry-driven capstone course on game programming: Extended abstract. *SIGCSE '05: Proceedings of the 36th SIGCSE Technical Symposium on Computer Science Education,* St. Louis, MO, United States, 91-95. Retrieved September 4, 2006 from http://doi.acm.org/10.1145/1047344.1047387

Parker, J. R., & Becker, K. (2003). Measuring effectiveness of constructivist and behaviourist assignments in CS102. *ITiCSE '03: Proceedings of the 8th Annual Conference on Innovation and Technology in Computer Science Education,* Thessaloniki, Greece, 40-44. Retrieved September 4, 2006 from http://doi.acm.org/10.1145/961511.961526

Parkinson, A., & Redmond, J. A. (2002). Do cognitive styles affect learning performance in different computer media? *ITiCSE '02: Proceedings of the 7th Annual Conference on Innovation and Technology in Computer Science Education,* Aarhus, Denmark, 39-43. Retrieved September 4, 2006 from http://doi.acm.org/10.1145/544414.544427

Parkinson, A., Redmond, J. A., & Walsh, C. (2004). Accommodating field-dependence: A cross-over study. *ITiCSE '04: Proceedings of the 9th Annual SIGCSE Conference on Innovation and Technology in Computer Science Education,* Leeds, United Kingdom, 72-76. Retrieved September 4, 2006 from http://doi.acm.org/10.1145/1007996.1008018

Parrish, A., Dixon, B., & Cordes, D. (2001). Binary software components in the undergraduate computer science curriculum. *SIGCSE '01: Proceedings of the Thirty-Second SIGCSE Technical Symposium on Computer Science Education,* Charlotte, NC, United States, 332-336. Retrieved September 4, 2006 from http://doi.acm.org/10.1145/364447.364615

Pears, A. & Olsson, H. (2004). Explanograms: low overhead multi-media learning resources. *Kolin Kolistelut – Koli Calling 2004, Proceedings of the Fourth Annual Finnish Annual Finnish Baltic Sea Conference on Computer Science Education,* Koli, Finland, 67-74. Retrieved September 4, 2006 from http://cs.joensuu.fi/kolistelut/archive/2004/koli_proc_2004.pdf

Perrenet, J., Groote, J. F., & Kaasenbrood, E. (2005). Exploring students' understanding of the concept of algorithm: Levels of abstraction. *ITiCSE '05: Proceedings of the 10th Annual SIGCSE Conference on Innovation and Technology in Computer Science Education,* Caparica, Portugal, 64-68. Retrieved September 4, 2006 from http://doi.acm.org/10.1145/1067445.1067467

Phoha, V. V. (2001). An interactive dynamic model for integrating knowledge management methods and knowledge sharing technology in a traditional classroom. *SIGCSE '01: Proceedings of the Thirty-Second SIGCSE Technical Symposium on Computer Science Education,* Charlotte, NC, United States, 144-148. Retrieved September 4, 2006 from http://doi.acm.org/10.1145/364447.364567

Pillay, N. (2004). A first course in genetic programming. *ITiCSE-WGR '04: Working Group Reports from ITiCSE on Innovation and Technology in Computer Science Education,* Leeds, United Kingdom, 93-96. Retrieved September 4, 2006 from http://doi.acm.org/10.1145/1044550.1041667

Pillay, N. (2003). Developing intelligent programming tutors for novice programmers. *SIGCSE Bulletin (Association for Computing Machinery, Special Interest Group on Computer Science Education), 35*(2), 78-82.

Pollock, L., McCoy, K., Carberry, S., Hundigopal, N., & You, X. (2004). Increasing high school girls' self confidence and awareness of CS through a positive summer experience. *SIGCSE '04: Proceedings of the 35th SIGCSE Technical Symposium on Computer Science Education,* Norfolk, VA, United States, 185-189. Retrieved September 4, 2006 from http://doi.acm.org/10.1145/971300.971369

Preston, J. A., & Wilson, L. (2001). Offering CS1 on-line reducing campus resource demand while improving the learning environment. *SIGCSE '01: Proceedings of the Thirty-Second SIGCSE Technical Symposium on Computer Science Education,* Charlotte, NC, United States, 342-346. Retrieved September 4, 2006 from http://doi.acm.org/10.1145/364447.364618

Prior, J. C., & Lister, R. (2004). The backwash effect on SQL skills grading. *ITiCSE '04: Proceedings of the 9th Annual SIGCSE Conference on Innovation and Technology in Computer Science Education,* Leeds, United Kingdom, 32-36. Retrieved September 4, 2006 from http://doi.acm.org/10.1145/1007996.1008008

Proulx, V. K. (2000). Programming patterns and design patterns in the introductory computer science course. *SIGCSE '00: Proceedings of the Thirty-First SIGCSE Technical Symposium on Computer Science Education,* Austin, TX, United States, 80-84. Retrieved September 4, 2006 from http://doi.acm.org/10.1145/330908.331819

Pullen, J. M., Norris, E., & Fix, M. (2000). Teaching C++ in a multi-user virtual environment. *SIGCSE Bulletin (Association for Computing Machinery, Special Interest Group on Computer Science Education), 32*(2), 60-64.

Rajaravivarma, R. (2005). A games-based approach for teaching the introductory programming course. *SIGCSE Bulletin (Association for Computing Machinery, Special Interest Group on Computer Science Education), 37*(4), 98-102.

Ramakrishnan, S., & Nwosu, E. (2003). DBMS course: Web based database administration tool and class projects. *SIGCSE '03: Proceedings of the 34th SIGCSE Technical Symposium on Computer Science Education,* Reno, NV, United States, 16-20. Retrieved September 4, 2006 from http://doi.acm.org/10.1145/611892.611922

Randall, C., Price, B., & Reichgelt, H. (2003). Women in computing programs: Does the incredible shrinking pipeline apply to all computing programs? *SIGCSE Bulletin (Association for Computing Machinery, Special Interest Group on Computer Science Education), 35*(4), 55-59.

Randolph, J. J., Bednarik, R. & Myller, N. (2005). A methodological review of the articles published in the proceedings of Koli Calling 2001-2004. *Kolin Kolistelut – Koli Calling 2005, Proceedings of the Fourth Annual Finnish Annual Finnish Baltic Sea Conference on Computer Science Education,* Koli, Finland, 103-111. Retrieved September 4, 2006 from http://cs.joensuu.fi/kolistelut/archive/2005/koli_proc_2005.pdf

Raner, M. (2000). Teaching object-orientation with the object visualization and annotation language (OVAL). *ITiCSE '00: Proceedings of the 5th Annual SIGCSE/SIGCUE ITiCSE conference on Innovation and Technology in Computer Science Education,* Helsinki, Finland, 45-48. Retrieved September 4, 2006 from http://doi.acm.org/10.1145/343048.343071

Rasala, R. (2000). Toolkits in first year computer science: A pedagogical imperative. *SIGCSE '00: Proceedings of the Thirty-First SIGCSE Technical Symposium on Computer Science Education,* Austin, TX, United States, 185-191. Retrieved September 4, 2006 from http://doi.acm.org/10.1145/330908.331852

Rasala, R. (2003). Embryonic object versus mature object: Object-oriented style and pedagogical theme. *ITiCSE '03: Proceedings of the 8th Annual Conference on Innovation and Technology in Computer Science Education,* Thessaloniki, Greece, 89-93, Retrieved September 4, 2006 from http://doi.acm.org/10.1145/961511.961538

Rasala, R., Raab, J., & Proulx, V. K. (2001). Java power tools: Model software for teaching object-oriented design. *SIGCSE '01: Proceedings of the Thirty-Second SIGCSE Technical Symposium on Computer Science Education,* Charlotte, NC, United States, 297-301. Retrieved September 4, 2006 from http://doi.acm.org/10.1145/364447.364606

Redmond, J. A., Walsh, C., & Parkinson, A. (2003). Equilibriating instructional media for cognitive styles. *ITiCSE '03: Proceedings of the 8th Annual Conference on Innovation and Technology in Computer Science Education,* Thessaloniki, Greece, 55-59. Retrieved September 4, 2006 from http://doi.acm.org/10.1145/961511.961529

Reges, S. (2002). Can C\# replace java in CS1 and CS2? *ITiCSE '02: Proceedings of the 7th Annual Conference on Innovation and Technology in Computer Science Education,* Aarhus, Denmark, 4-8. Retrieved September 4, 2006 from http://doi.acm.org/10.1145/544414.544419

Reges, S. (2000). Conservatively radical java in CS1. *SIGCSE '00: Proceedings of the Thirty-First SIGCSE Technical Symposium on Computer Science Education,* Austin, TX, United States, 85-89. Retrieved September 4, 2006 from http://doi.acm.org/10.1145/330908.331821

Reis, C., & Cartwright, R. (2004). Taming a professional IDE for the classroom. *SIGCSE '04: Proceedings of the 35th SIGCSE Technical Symposium on Computer Science Education,* Norfolk, VA, United States, 156-160. Retrieved September 4, 2006 from http://doi.acm.org/10.1145/971300.971357

Richards, B. (2000). Bugs as features: Teaching network protocols through debugging. *SIGCSE '00: Proceedings of the Thirty-First SIGCSE Technical Symposium on Computer Science Education,* Austin, TX, United States, 256-259. Retrieved September 4, 2006 from http://doi.acm.org/10.1145/330908.331865

Richards, B., & Stull, B. (2004). Teaching wireless networking with limited resources. *SIGCSE '04: Proceedings of the 35th SIGCSE Technical Symposium on Computer Science Education,* Norfolk, VA, United States, 306-310. Retrieved September 4, 2006 from http://doi.acm.org/10.1145/971300.971408

Robbert, M. A. (2000). Enhancing the value of a project in the database course. *SIGCSE '00: Proceedings of the Thirty-First SIGCSE Technical Symposium on Computer Science Education,* Austin, TX, United States, 36-40. Retrieved September 4, 2006 from http://doi.acm.org/10.1145/330908.331807

Robbins, S. (2003). Using remote logging for teaching concurrency. *SIGCSE '03: Proceedings of the 34th SIGCSE Technical Symposium on Computer Science Education,* Reno, NV, United States, 177-181. Retrieved September 4, 2006 from http://doi.acm.org/10.1145/611892.611963

Robbins, S. (2001). Starving philosophers: Experimentation with monitor synchronization. *SIGCSE '01: Proceedings of the Thirty-Second SIGCSE Technical Symposium on Computer Science Education,* Charlotte, NC, United States, 317-321. Retrieved September 4, 2006 from http://doi.acm.org/10.1145/364447.364612

Robins, A., Roundtree, J., & Roundtree, N. (2003). Learning and teaching programming: A review and discussion. *Computer Science Education, 13*(2), 137-172.

Roberts, E. S., Kassianidou, M., & Irani, L. (2002). Encouraging women in computer science. *SIGCSE Bulletin (Association for Computing Machinery, Special Interest Group on Computer Science Education), 34*(2), 84-88.

Rodger, S. H. (2002). Introducing computer science through animation and virtual worlds. *SIGCSE '02: Proceedings of the 33rd SIGCSE Technical Symposium on Computer Science Education,* Cincinnati, OH, United States 186-190. Retrieved September 4, 2006 from http://doi.acm.org/10.1145/563340.563411

Rosbottom, J., Crellin, J., & Fysh, D. (2000). A generic model for on-line learning. *ITiCSE '00: Proceedings of the 5th Annual SIGCSE/SIGCUE ITiCSE conference on Innovation and Technology in Computer Science Education,* Helsinki, Finland. 108-111. Retrieved September 4, 2006 from http://doi.acm.org/10.1145/343048.343131

Rosson, M. B., Carroll, J. M., & Rodi, C. M. (2004). Case studies for teaching United Statesbility engineering. *SIGCSE '04: Proceedings of the 35th SIGCSE Technical Symposium on Computer Science Education,* Norfolk, VA, United States, 36-40. Retrieved September 4, 2006 from http://doi.acm.org/10.1145/971300.971315

Rountree, J., Rountree, N., Robins, A., & Hannah, R. (2005}). Observations of student competency in a CS1 course. *Seventh Australasian Computing Education Conference (ACE2005); Conferences in Research and Practice in Information Technology, 42,* 145-149.

Rößling, G., & Naps, T. L. (2002). A testbed for pedagogical requirements in algorithm visualizations. *ITiCSE '02: Proceedings of the 7th Annual Conference on Innovation and Technology in Computer Science Education,* Aarhus, Denmark, 96-100. Retrieved September 4, 2006 from http://doi.acm.org/10.1145/544414.544446

Saastamoinen, K. (2003). Distance learning experiences from the mathematical modeling course. *Kolin Kolistelut – Koli Calling 2003, Proceedings of the Third Annual Finnish Annual Finnish Baltic Sea Conference on Computer Science Education,* Koli, Finland, 22-26. Retrieved September 4, 2006 from http://cs.joensuu.fi/kolistelut/archive/2003/koli_proc_2003.pdf

Sajaniemi, J., & Kuittinen, M. (2005). An experiment on using roles of variables in teaching introductory programming. *Computer Science Education, 15*(1), 59-82.

Sanders, I. (2002). Teaching empirical analysis of algorithms. *SIGCSE '02: Proceedings of the 33rd SIGCSE Technical Symposium on Computer Science Education,* Cincinnati, OH, United States 321-325. Retrieved September 4, 2006 from http://doi.acm.org/10.1145/563340.563468

Sanders, D. (2002). Extreme programming: The student view. *Computer Science Education*, *12*(3), 235-250.

Sanders, K. E., & McCartney, R. (2003). Program assessment tools in computer science: A report from the trenches. *SIGCSE '03: Proceedings of the 34th SIGCSE Technical Symposium on Computer Science Education*, Reno, NV, United States, 31-35. Retrieved September 4, 2006 from http://doi.acm.org/10.1145/611892.611926

Sanders, K., & McCartney, R. (2004). Collected wisdom: Assessment tools for computer science programs. *Computer Science Education*, *14*(3), 183-203.

Sarkar, N. I., & Craig, T. M. (2004). Illustrating computer hardware concepts using PIC-based projects. *SIGCSE '04: Proceedings of the 35th SIGCSE Technical Symposium on Computer Science Education*, Norfolk, VA, United States, 270-274. Retrieved September 4, 2006 from http://doi.acm.org/10.1145/971300.971395

Saxon, C. S. (2003). Object-oriented recursive descent parsing in C\#. *SIGCSE Bulletin (Association for Computing Machinery, Special Interest Group on Computer Science Education)*, *35*(4), 82-85.

Schaub, S. (2000). Teaching java with graphics in CS1. *SIGCSE Bulletin (Association for Computing Machinery, Special Interest Group on Computer Science Education)*, *32*(2), 71-73.

Schulte, C. (2002). A picture is worth a thousand words? – UML as a multimedia learning tool. *Kolin Kolistelut – Koli Calling 2002, Proceedings of the Second Annual Finnish Annual Finnish Baltic Sea Conference on Computer Science Education*, Koli, Finland, 57-62. Retrieved September 4, 2006 from http://cs.joensuu.fi/kolistelut/archive/2002/koli_proc_2002.pdf

Schulte, C. (2005). Dynamic object structures as a conceptual framework for teaching object oriented concepts to novices. *Kolin Kolistelut – Koli Calling 2005, Proceedings of the Fourth Annual Finnish Annual Finnish Baltic Sea Conference on Computer Science Education*, Koli, Finland, 113-118. Retrieved September 4, 2006 from http://cs.joensuu.fi/kolistelut/archive/2005/koli_proc_2005.pdf

Seppälä, O., Malmi, L. & Korhonen, A. (2005). Observations on student errors in algorithm simulation exercises. *Kolin Kolistelut – Koli Calling 2005, Proceedings of the Fourth Annual Finnish Annual Finnish Baltic Sea Conference on Computer Science Education*, Koli, Finland, 81-86. Retrieved September 4, 2006 from http://cs.joensuu.fi/kolistelut/archive/2005/koli_proc_2005.pdf

Shaffer, C. A. (2004). Buffer pools and file processing projects for an undergraduate data structures course. *SIGCSE '04: Proceedings of the 35th SIGCSE Technical Symposium on Computer Science Education*, Norfolk, VA, United States, 175-178. Retrieved September 4, 2006 from http://doi.acm.org/10.1145/971300.971362

Shaw, K., & Dermoudy, J. (2005}). Engendering an empathy for software engineering. *Seventh Australasian Computing Education Conference (ACE2005); Conferences in Research and Practice in Information Technology, 42,* 135-144.

Sheard, J. (2004). Electronic learning communities: Strategies for establishment and management. *ITiCSE '04: Proceedings of the 9th Annual SIGCSE Conference on Innovation and Technology in Computer Science Education,* Leeds, United Kingdom, 37-41. Retrieved September 4, 2006 from http://doi.acm.org/10.1145/1007996.1008009

Sheard, J., & Dick, M. (2003). Influences on cheating practice of graduate students in IT courses: What are the factors? *ITiCSE '03: Proceedings of the 8th Annual Conference on Innovation and Technology in Computer Science Education,* Thessaloniki, Greece, 45-49. Retrieved September 4, 2006 from http://doi.acm.org/10.1145/961511.961527

Simmonds, A. (2003}). Student learning experience with an industry certification course at university. *Fifth Australasian Computing Education Conference (ACE2003); Conferences in Research and Practice in Information Technology, 20,* 143-147.

Simon, B., Anderson, R., Hoyer, C., & Su, J. (2004). Preliminary experiences with a tablet PC based system to support active learning in computer science courses. *ITiCSE '04: Proceedings of the 9th Annual SIGCSE Conference on Innovation and Technology in Computer Science Education,* Leeds, United Kingdom, 213-217. Retrieved September 4, 2006 from http://doi.acm.org/10.1145/1007996.1008053

Smarkusky, D., Dempsey, R., Ludka, J., & Quillettes, F. d. (2005). Enhancing team knowledge: Instruction vs. experience. *SIGCSE '05: Proceedings of the 36th SIGCSE Technical Symposium on Computer Science Education,* St. Louis, MO, United States, 460-464. Retrieved September 4, 2006 from http://doi.acm.org/10.1145/1047344.1047493

Soh, L., Samal, A., Person, S., Nugent, G., & Lang, J. (2005). Analyzing relationships between closed labs and course activities in CS1. *ITiCSE '05: Proceedings of the 10th Annual SIGCSE Conference on Innovation and Technology in Computer Science Education,* Caparica, Portugal, 183-187. Retrieved September 4, 2006 from http://doi.acm.org/10.1145/1067445.1067497

Soh, L., Samal, A., Person, S., Nugent, G., & Lang, J. (2005). Designing, implementing, and analyzing a placement test for introductory CS courses. *SIGCSE '05: Proceedings of the 36th SIGCSE Technical Symposium on Computer Science Education,* St. Louis, MO, United States, 505-509. Retrieved September 4, 2006 from http://doi.acm.org/10.1145/1047344.1047504

Solomon, A. (2003}). Applying NAILS to blackboard. *Fifth Australasian Computing Education Conference (ACE2003); Conferences in Research and Practice in Information Technology, 20,* 263-266.

Sooriamurthi, R. (2001). Problems in comprehending recursion and suggested solutions. *ITiCSE '01: Proceedings of the 6th Annual Conference on Innovation and Technology in Computer Science Education,* Canterbury, United Kingdom, 25-28. Retrieved September 4, 2006 from http://doi.acm.org/10.1145/377435.377458

Spooner, D. L. (2000). A bachelor of science in information technology: An interdisciplinary approach. *SIGCSE '00: Proceedings of the Thirty-First SIGCSE Technical Symposium on Computer Science Education,* Austin, TX, United States, 285-289. Retrieved September 4, 2006 from http://doi.acm.org/10.1145/330908.331871

Steenkiste, P. (2003). A network project course based on network processors. *SIGCSE '03: Proceedings of the 34th SIGCSE Technical Symposium on Computer Science Education,* Reno, NV, United States, 262-266. Retrieved September 4, 2006 from http://doi.acm.org/10.1145/611892.611984

Stern, L., & Naish, L. (2002). Visual representations for recursive algorithms. *SIGCSE '02: Proceedings of the 33rd SIGCSE Technical Symposium on Computer Science Education,* Cincinnati, OH, United States 196-200. Retrieved September 4, 2006 from http://doi.acm.org/10.1145/563340.563414

Stevenson, D. E., & Phillips, A. T. (2003). Implementing object equivalence in java using the template method design pattern. *SIGCSE '03: Proceedings of the 34th SIGCSE Technical Symposium on Computer Science Education,* Reno, NV, United States, 278-282. Retrieved September 4, 2006 from http://doi.acm.org/10.1145/611892.611987

Sweedyk, E., & Keller, R. M. (2005). Fun and games: A new software engineering course. *ITiCSE '05: Proceedings of the 10th Annual SIGCSE Conference on Innovation and Technology in Computer Science Education,* Caparica, Portugal, 138-142. Retrieved September 4, 2006 from http://doi.acm.org/10.1145/1067445.1067485

Tegos, G. K., Stoyanova, D. V., & Onkov, K. Z. (2005). E-learning of trend modeling in a web-environment. *SIGCSE Bulletin (Association for Computing Machinery, Special Interest Group on Computer Science Education), 37*(2), 70-74.

Tenenberg, J. (2003). A framework approach to teaching data structures. *SIGCSE '03: Proceedings of the 34th SIGCSE Technical Symposium on Computer Science Education,* Reno, NV, United States, 210-214. Retrieved September 4, 2006 from http://doi.acm.org/10.1145/611892.611971

Thomas, P. (2003). The evaluation of electronic marking of examinations. *ITiCSE '03: Proceedings of the 8th Annual Conference on Innovation and Technology in Computer Science Education,* Thessaloniki, Greece, 50-54. Retrieved September 4, 2006 from http://doi.acm.org/10.1145/961511.961528

Thomas, P., Waugh, K., & Smith, N. (2005). Experiments in the automatic marking of ER-diagrams. *ITiCSE '05: Proceedings of the 10th Annual SIGCSE Conference on Innovation and Technology in Computer Science Education,* Caparica, Portugal, 158-162. Retrieved September 4, 2006 from http://doi.acm.org/10.1145/1067445.1067490

Tikvati, A., Ben-Ari, M., & Kolikant, Y. B. (2004). Virtual trees for the Byzantine generals algorithm. *SIGCSE '04: Proceedings of the 35th SIGCSE Technical Symposium on Computer Science Education,* Norfolk, VA, United States, 392-396. Retrieved September 4, 2006 from http://doi.acm.org/10.1145/971300.971435

Tolhurst, D., & Baker, B. (2003}). A new approach to a first year undergraduate information systems course. *Fifth Australasian Computing Education Conference (ACE2003); Conferences in Research and Practice in Information Technology, 20,* 169-177.

Townsend, G. C. (2002). People who make a difference: Mentors and role models. *SIGCSE Bulletin (Association for Computing Machinery, Special Interest Group on Computer Science Education), 34*(2), 57-61.

Traynor, C., & McKenna, M. (2003). Service learning models connecting computer science to the community. *SIGCSE Bulletin (Association for Computing Machinery, Special Interest Group on Computer Science Education), 35*(4), 43-46.

Treu, K. (2002). To teach the unteachable class: An experimental course in web-based application design. *SIGCSE '02: Proceedings of the 33rd SIGCSE Technical Symposium on Computer Science Education,* Cincinnati, OH, United States 201-205. Retrieved September 4, 2006 from http://doi.acm.org/10.1145/563340.563416

Truong, N., Roe, P., & Bancroft, P. (2004}). Static analysis of students' java programs. *Sixth Australasian Computing Education Conference (ACE2004); Conferences in Research and Practice in Information Technology, 30,* 317-325.

Tuttle, S. M. (2000). A capstone course for a computer information systems major. *SIGCSE '00: Proceedings of the Thirty-First SIGCSE Technical Symposium on Computer Science Education,* Austin, TX, United States, 265-269. Retrieved September 4, 2006 from http://doi.acm.org/10.1145/330908.331867

VanDeGrift, T. (2004). Coupling pair programming and writing: Learning about students' perceptions and processes. *SIGCSE '04: Proceedings of the 35th SIGCSE Technical Symposium on Computer Science Education,* Norfolk, VA, United States, 2-6. Retrieved September 4, 2006 from http://doi.acm.org/10.1145/971300.971306

Vandenberg, S., & Wollowski, M. (2000). Introducing computer science using a breadth-first approach and functional programming. *SIGCSE '00: Proceedings of the Thirty-First SIGCSE*

Technical Symposium on Computer Science Education, Austin, TX, United States, 180-184. Retrieved
    September 4, 2006 from http://doi.acm.org/10.1145/330908.331851

van der Veer, G., & van Vliet, H. (2003). A plea for a poor man's HCI component in software
    engineering and computer science curricula; after all: The human-computer interface is the
    system. *Computer Science Education, 13*(3), 207-225.

Vaughn, J. (2001). Teaching industrial practices in an undergraduate software engineering course.
    Computer Science Education, 11(1), 21-32.

Veen, M. V., Mulder, F., & Lemmen, K. (2004). What is lacking in curriculum schemes for
    computing/informatics? *ITiCSE '04: Proceedings of the 9th Annual SIGCSE Conference on Innovation
    and Technology in Computer Science Education,* Leeds, United Kingdom, 186-190. Retrieved
    September 4, 2006 from http://doi.acm.org/10.1145/1007996.1008046

Velázquez-Iturbide, J. (2000). Recursion in gradual steps (is recursion really that difficult?).
    *SIGCSE '00: Proceedings of the Thirty-First SIGCSE Technical Symposium on Computer Science
    Education,* Austin, TX, United States, 310-314. Retrieved September 4, 2006 from
    http://doi.acm.org/10.1145/330908.331876

Venables, A., & Haywood, L. (2003). Programming students need instant feedback! *Fifth
    Australasian Computing Education Conference (ACE2003); Conferences in Research and Practice in
    Information Technology, 20,* 267-272.

Ventura, P., & Ramamurthy, B. (2004). Wanted: CS1 students. no experience required. *SIGCSE
    '04: Proceedings of the 35th SIGCSE Technical Symposium on Computer Science Education,* Norfolk,
    VA, United States, 240-244. Retrieved September 4, 2006 from
    http://doi.acm.org/10.1145/971300.971387

Vidal, J., M., & Buhler, P. (2002). Using RoboCup to teach multiagent systems and the distributed
    mindset. *SIGCSE '02: Proceedings of the 33rd SIGCSE Technical Symposium on Computer Science
    Education,* Cincinnati, OH, United States 3-7. Retrieved September 4, 2006 from
    http://doi.acm.org/10.1145/563340.563344

von Konsky, B. R., Ivins, J., & Robey, M. (2005}). Using PSP to evaluate student effort in achieving
    learning - outcomes in a software engineering assignment. *Seventh Australasian Computing
    Education Conference (ACE2005); Conferences in Research and Practice in Information Technology, 42,*
    193-201.

Wagner, P. J., Shoop, E., & Carlis, J. V. (2003). Using scientific data to teach a database systems
    course. *SIGCSE '03: Proceedings of the 34th SIGCSE Technical Symposium on Computer Science
    Education,* Reno, NV, United States, 224-228. Retrieved September 4, 2006 from
    http://doi.acm.org/10.1145/611892.611975

Warren, P. (2004}). Learning to program: Spreadsheets, scripting and HCI. *Sixth Australasian Computing Education Conference (ACE2004); Conferences in Research and Practice in Information Technology, 30,* 327-333.

Wells, M. A., & Brook, P. W. (2004}). Conversational KM - student driven learning. *Sixth Australasian Computing Education Conference (ACE2004); Conferences in Research and Practice in Information Technology, 30,* 335-341.

Whiddett, R., Jackson, B., & Handy, J. (2000). Teaching information systems management skills: Using integrated projects and case studies. *Computer Science Education, 10*(2), 165-177.

White, K. (2003). A comprehensive CMPS II semester project. *SIGCSE Bulletin (Association for Computing Machinery, Special Interest Group on Computer Science Education), 35*(2), 70-73.

Wicentowski, R., & Newhall, T. (2005). Using image processing projects to teach CS1 topics. *SIGCSE '05: Proceedings of the 36th SIGCSE Technical Symposium on Computer Science Education,* St. Louis, MO, United States, 287-291. Retrieved September 4, 2006 from http://doi.acm.org/10.1145/1047344.1047445

Wiedenbeck, S. (2005). Factors affecting the success of non-majors in learning to program. *ICER '05: Proceedings of the 2005 International Workshop on Computing Education Research,* Seattle, WA, United States, 13-24. Retrieved September 4, 2006 from http://doi.acm.org/10.1145/1089786.1089788

Wilkerson, M., Griswold, W. G., & Simon, B. (2005). Ubiquitous presenter: Increasing student access and control in a digital lecturing environment. *SIGCSE '05: Proceedings of the 36th SIGCSE Technical Symposium on Computer Science Education,* St. Louis, MO, United States, 116-120. Retrieved September 4, 2006 from http://doi.acm.org/10.1145/1047344.1047394

Williams, L., Wiebe, E., Yang, K., Ferzli, M., & Miller, C. (2002). In support of pair programming in the introductory computer science course. *Computer Science Education, 12*(3), 197-212.

Wiseman, Y. (2005). Advanced non-distributed operating systems course. *SIGCSE Bulletin (Association for Computing Machinery, Special Interest Group on Computer Science Education), 37*(2), 65-69.

Wolz, U. (2001). Teaching design and project management with lego RCX robots. *SIGCSE '01: Proceedings of the Thirty-Second SIGCSE Technical Symposium on Computer Science Education,* Charlotte, NC, United States, 95-99. Retrieved September 4, 2006 from http://doi.acm.org/10.1145/364447.364551

Wong, Y., Burg, J., & Strokanova, V. (2004). Digital media in computer science curricula. *SIGCSE '04: Proceedings of the 35th SIGCSE Technical Symposium on Computer Science Education,* Norfolk,

VA, United States, 427-431. Retrieved September 4, 2006 from
http://doi.acm.org/10.1145/971300.971444

Woodford, K., & Bancroft, P. (2005}). Multiple choice questions not considered harmful. *Seventh Australasian Computing Education Conference (ACE2005); Conferences in Research and Practice in Information Technology, 42* 109-116.

Yacef, K. (2004}). Making large class teaching more adaptive with the logic-ITA. *Sixth Australasian Computing Education Conference (ACE2004); Conferences in Research and Practice in Information Technology, 30,* 343-347.

Yehezkel, C. (2002). A taxonomy of computer architecture visualizations. *ITiCSE '02: Proceedings of the 7th Annual Conference on Innovation and Technology in Computer Science Education,* Aarhus, Denmark, 101-105. Retrieved September 4, 2006 from
http://doi.acm.org/10.1145/544414.544447

Yoo, S., & Hovis, S. (2004). Remote access internetworking laboratory. *SIGCSE '04: Proceedings of the 35th SIGCSE Technical Symposium on Computer Science Education,* Norfolk, VA, United States, 311-314. Retrieved September 4, 2006 from http://doi.acm.org/10.1145/971300.971409

Yurcik, W., & Brumbaugh, L. (2001). A web-based little man computer simulator. *SIGCSE '01: Proceedings of the Thirty-Second SIGCSE Technical Symposium on Computer Science Education,* Charlotte, NC, United States, 204-208. Retrieved September 4, 2006 from
http://doi.acm.org/10.1145/364447.364585

Zaring, A. (2001). CCL: A subject language for compiler projects. *Computer Science Education, 11*(2), 135-163.

Zelle, J. M., & Figura, C. (2004). Simple, low-cost stereographics: VR for everyone. *SIGCSE '04: Proceedings of the 35th SIGCSE Technical Symposium on Computer Science Education,* Norfolk, VA, United States, 348-352. Retrieved September 4, 2006 from
http://doi.acm.org/10.1145/971300.971421

Zeller, A. (2000). Making students read and review code. *ITiCSE '00: Proceedings of the 5th Annual SIGCSE/SIGCUE ITiCSE conference on Innovation and Technology in Computer Science Education,* Helsinki, Finland. 89-92. Retrieved September 4, 2006 from
http://doi.acm.org/10.1145/343048.343090

APPENDIX B

Methodological Review Coding Form

DE0 = _ _ _

DE00=_ _

DE000. 1=yes, 2=no.

DE1. (reviewer): 1 = Justus, 2 = Roman, 3 = Nikko, 4 =other _____

DE2. (forum): 1 = SICGSE proceedings, 2 = SIGCSE bulletin, 3 = ITICES, 4=CSER, 5=KOLI, 6=ICER, 7 = JCSE, 8 = ACE.

DE3. (year): 0=2000, 1=2001, 2=2002, 3=2003, 4=2004, 5=2005.

DE4. (volume) _ _ _ (three numerical digits – use zero for blank digits; e.g., Volume 1 would be 001.)

DE5. (issue) _ _ (two numerical digits)

DE6. (page) _ _ _ _ (up to four digits)

DE6a. (pages) _ _ _ _

DE7. (region) 1 = Africa, 2 = Asian-Pacific or Eurasia, 3 = Europe, 4 = Middle East, 5 = North America, 6 = South or Central America, 7 = IMPDET

DE7a (university) Write in. _____

DE7b (authors) # _____

DE7c (name) Last name, Initials _____ __ __

DE8. (Subject) 1 = New way to organize a course, 2 = Tool, 3 = Teaching programming language category, 4 = Curriculum, 5= Visualization, 6 = Simulation, 7 = Parallel computing, 8 = Other.

DE8a (Valentine) 1= Experimental, 2=Marco Polo, 3= Tools, 4= John Henry, 5= Philosophy 6= Nifty

DE9. (human participants) 1 = yes, 2 = no. (If yes, go to DE9a ; if no go to A9.)

DE9a (anecdotal) 1=yes, 2=no.
(if yes, go to M21.)

_____

Type of Papers that Did Not Report Research on Human Subjects

A9. (type of other) 1 = Literature review, 2 = Program description, 3 = Theory, Methodology, Philosophy paper, 4 = Technical investigation, 5 = Other (if 1-4, end; if 5 go to A10)

A10 (Other other) Write in a short description (End).

_____

Methodology Type

M21. Experimental/quasi-experimental 1 = yes, 2 = no
(If M21=yes, go to AS5, else go to M22.)

    AS5. (assignment) 1 = self-selection 2= random 3 = researcher-assigned

M22. Explanatory descriptive   1 = yes, 2 = no

M23. Exploratory description   1 = yes, 2 = no

M24. Correlational   1 = yes, 2 = no

M25. Causal-comparative   1 = yes, 2 = no

M26. IMPDET or anecdotal   1 = yes, 2 = no

M27. (selection) 1 = random, 2 = intentional, 3 = convenience/preexisting

[Go to A11]

_____

Report Structure

A11. Abstract   1= narrative, 2 = structured, 3 = no abstract

A12. (introduce problem)   1 = yes, 2 = no

A13. (literature review)   1 = yes, 2 = no

A14. (purpose/rationale)   1 = yes, 2 = no

A15. (questions/hypotheses)   1 = yes, 2 = no

A16. (participants)   1 = yes, 2 = no

A16a (grade level)   1= preschool

                 2= k-3

                 3= 4-6

                 4= 7-9

                 5= 10-12

                 6=bachelor

                 7=masters

                 8=doctoral

                 9=post-doctoral

                 10=other

                 11=can't determine

A16b (Undergraduate

curriculum year)   1 = first year

                 2 = second year

                 3 = third year

                 4 = fourth year

A17. (settings)   1 = yes, 2 = no
A18. (instruments)   1 = yes, 2 = no, -9 = n/a
A19. (procedure)   1 = yes, 2 = no
A20. (results and discussion)   1 = yes, 2 = no
[Go to RD1, if M21 = 1, else go to I1.]

_____

Experimental Research Designs

RD1. (design) Was M21, marked as Yes 1 = yes, 2 = no
[if yes, RD2; If no go to I1]
RD2 (postonly) posttest, no controls   1 = yes, 2 = no
RD3 (post control) posttest, with controls,   1 = yes, 2 = no
RD4 (prepost only= pretest/posttest without controls 1 = yes, 2 = no
RD5 (prepost control) pretest/posttest with controls   1 = yes, 2 = no
RD6 (repeated) group repeated measures   1 = yes, 2 = no
RD7 (multiple) multiple factor   1 = yes, 2 = no

RD11 (factor?) If group repeated measures,
        was there an experimental between group factor?   1= yes, 2 = no

RD8 (single) single-subject   1 = yes, 2 = no
RD9 (other) other   1 = yes, 2 = no
[if RD9, go to RD10]
RD 10 (explain) If other, explain
RDH (posttest only highest)   1 = yes, 2 = no

_____

Independent Variables (interventions)

I1. Was an independent (manipulatable) variable used in this study? 1 = yes, 2 = no
[If yes got to I2, if no go to D1]
I2 (student instruction) 1 = yes, 2 = no

I3 (teacher instruction)   1 = yes, 2 = no

I4 (CS fair /contest)   1 = yes, 2 = no

I5 (mentoring)   1 = yes, 2 = no

I6 (Speakers at school)   1 = yes, 2 = no

I7 (CS field trips)   1 = yes, 2 = no

I8 (other)   1 = yes, 2 = no

If I8a (explain) If other, explain:

[Go to D1]

Dependent Variables

D1 (attitudes)   1 = yes, 2 = no

D2 (attendance)   1 = yes, 2 = no

D3 (core achievement)   1 = yes, 2 = no

D4 (CS achievement)   1 = yes, 2 = no

D5 (teaching practices)   1 = yes, 2 = no

D6 (intentions for future)   1 = yes, 2 = no

D7 (program implementation)   1 = yes, 2 = no

D8 (costs and benefits $)   1 = yes, 2 = no

D9 (socialization)   1 = yes, 2 = no

D10 (computer use)   1 = yes, 2 = no

D11 (other)   1 = yes, 2 = no

D11a (explain) If D11, explain

[Go to M1]

Measures

M1 (grades)   1 = yes, 2 = no

M2 (diary)   1 = yes, 2 = no

M3 (questionnaire)   1 = yes, 2 = no

M3a (ques. psych) 1 = yes, 2 = no

M4 (log files)   1 = yes, 2 = no

M5 (test)   1 = yes, 2 = no

M5a (test psych)   1 = yes, 2 = no

M6 (interviews)   1 = yes, 2 = no

M7 (direct)   1 = yes, 2 = no

M7a (direct psych)   1 = yes, 2 = no

M8 (stand. Test)   1 = yes, 2 = no

M8a (psych. Stand)   1 = yes, 2 = no

M9 (student work)   1 = yes, 2 = no

M10 (focus groups)   1 = yes, 2 = no

M11 (existing data)   1 = yes, 2 = no

M12 (other)   1 = yes, 2 = no

M12a (explain) If other, explain:

[Go to F1]

_____

Factors — (Non-manipulatable Variables)

F1 (nm factor?) Were any nonmanipulatable factors
      examined as covariates?   1 = yes, 2 = no

[If yes, go to F2; if no go to S1]

 F2 (gender)   1 = yes, 2 = no

F3 (aptitude)   1 = yes, 2 = no

F4 (race/ethic origin)   1 = yes, 2 = no

F5 (nationality)   1 = yes, 2 = no

F6 (disability)   1 = yes, 2 = no

F7 (SES)   1 = yes, 2 = no

F8 (other)   1 = yes, 2 = no

F8a (explain) If F8, then explain:

[Go to S1]

Statistical Practices

S1. (quant) Were quantitative results reported? 1 = yes, 2 = no
[If yes, go to S2; if no end.]

S2. (inf.stats) Were inferential statistics used? 1 = yes, 2 = no
[If yes, go to S3; Else go to S8]]

S3 (parametric) Parametric test of location used? 1 = yes, 2 = no
 [Is yes, go to s3a; else go to s4]

S3a (means) Were cell means and cell variances
     or cell means, mean square error
     and degrees of freedom reported? 1 = yes, 2 = no

S4 (multi) Were multivariate analyses used?   1 = yes, 2 = no
[Is yes, go to s4a; else go to s5]
   S4a (means) Were cell means reported?   1 = yes, 2 = no
   S4b (sizes) Were cell sample sizes reported?   1 = yes, 2 = no
   S4c (variance) Was pooled within variance or
         covariance matrix reported?   1 = yes, 2 = no

S5 (correlational) Were correlational analyses done?   1 = yes, 2 = no
[Is yes, go to s5a; else go to s6]
   S5a (size) Was sample size reported?   1 = yes, 2 = no
   S5b (matrix) Was variance – covariance,
         or correlation matrix reported ?   1 = yes, 2 = no
S6 (nonparametric) Were nonparametric analyses used?   1 = yes, 2 no
[Is yes, go to s6a; else go to s7]

   S6a (raw data) Were raw data summarized?   1 = yes, 2 = no

183

S7 (small sample) Were analyses for very small samples done?   1 = yes, 2 = no
[Is yes, go to s7a; else go to s8]

   S7a (entire data set) Was entire data set reported?   1 = yes, 2 = no

S8 (effect size) Was an effect size reported?   1 = yes, 2 = no

[If yes, go to S8a, else end.]

   S8a (raw diff.) Was there a difference in
            means, proportions, medians, etc., reported?   1 = yes, 2 = no

   S8aa (variability) Was a measure of dispersion reported if   1 = yes, 2 = no
          a mean was reported? If a mean was not reported, then -9

   S8b (SMD) Standardized mean difference effect size 1 = yes, 2 = no

   S8c (Corr.) Correlational effect size   1 = yes, 2 = no

   S8d (OR) Odds ratios   1 = yes, 2 = no

   S8e (odds) Odds   1 = yes, 2 = no

   S8f (RR) Relative risk   1 = yes, 2 = no

   S8h (other) Other   1 = yes, 2 = no

   S8i (explain) Explain other
[end]

This page intentionally left blank.

APPENDIX C

Methodological Review Coding Book

Note: Unless other wise specified, every cell of the coding datasheet must be filled in. Use -9 to specify that a variable is not applicable. Do not leave cells blank.

DEMOGRAPHIC CHARACTERISTICS
In the variables in this section, the demographic characteristics of each study are coded.

DE0. (case) This is the case number. It will be assigned by the primary coder.

DE00. (category) This variable corresponds with the first two digits of the case number. It refers to Table 5; the letter corresponds with the row (forum) and the number corresponds with the year.

DE000. (kappa) This specifies if this case was used for interrater reliability estimates. 1 = *yes*, 2 = *no*.

DE1. (reviewer) Circle the number that corresponds with your name. If your name is not on the list, choose *other* and write in your name. (Choose one.)

DE2. (forum) Circle the number of the forum in which the article was published. (*SIGCSE* = SIGCSE technical symposium, *Bulletin* = June or December issue of SIGCSE Bulletin, *ITiCSE* = Innovation and Technology in Computer Science Education Conference, *CSE* = Computer Science Education, *ICER* = International Computer Science Education Research Workshop, *JCSE* = Journal of Computer Science Education Online, *ACE* = Australasian Computing Education Conference.) (Choose one.)

DE2a. (type of forum). Choose 1 if the forum where the article was published is a journal (i.e., if the article was not meant to be presented at a conference and published in a peer-reviewed forum, or if the title of the forum includes the term *journal*.). Choose 2 if the forum where the article was published is a conference proceeding (i.e., it was meant to be published at a conference and may or may not have been peer-reviewed.) In this case, choose 1 if the article was published in the June or December issues of SIGCSE Bulletin, *Computer Science Education,* or the *Journal of Computer Science Education Online*, otherwise choose 2.

DE3. (year) Write in the year in which the article was published. 0=2000, 1=2001, 2=2002, 3=2003, 4=2004, 5=2005.

DE4. (volume) Write in the volume in which the article was published. Use three digits (e.g., volume 5 = 005.) If there was not a volume number, write in 000.

DE5. (issue) Write in the issue in which the article was published. Use two digits (e.g., issue 2 = 02.) If there was not an issue number, write in 00.

DE6. (page) Write in the page on which the article began. Use four digits (e.g., if the article began on page 347 = 0347.) If there was not a page number, write in 0000.

DE6a. (pages) Write in how many pages long the article was. If the article had no page numbers write in *-9.*

DE7. (region) Choose the region of origin of the first author's affiliation. Choose only one. If the regions of first author's affiliation cannot be determined, use 7 (IMPDET = impossible to determine). (This variable was derived from previous the methodological reviews: Randolph [2005, in press], Randolph, Bednarik, & Myller [2005] and Randolph, Bednarik, Silander, Lopez-Gonzales, Myller, & Sutinen [2005])

DE7a. (university) Write in the name of the university or affiliation of the first author.

DE7b. (authors) Write in the number of authors.

DE7c. (name) Write in the name of the first author. Last name first and then initials, which are followed by a period (e.g. Justus Joseph Randolph = Randolph, J. J.). Use a hyphen if a name is hyphenated (Randolph-Ratilainen), but do not use special characters.

TYPE OF PAPER

These variables group the papers into papers that did research on human participants and those that did not. For those that did not, they are further classified.

DE8. Subject of study. (This variable comes from a review of the subject matter discussed in *SIGCSE Bulletin* articles 1990-2004 [Kinnunen, n.d.]. They were derived using a emergent approach. Quotes are from Kinnunen, n.d.) Only choose one. If an article could belong to more than one category, choose the category that the article discusses the most. 'Tool' articles supersede 'new ways to teach a course,' when the new was to teach a course includes using a new tool.

- Choose 1 if the subject of the study involved new ways to organize a course. For example some courses might include "single new assignments" or "more drastic changes in the course." An example is Mattis (1995).
- Choose 2 if the article discusses "a new tool or experiences using a new tool." An example of a tool article is Dawson-Howe (1995)
- Choose 3 if the article discusses teaching programming languages. This includes articles that discuss "which language is best for students as a first language and papers that discuss about how some smaller section of a language should be taught." An example of this type of paper is Cole (1990).
- Choose 4 if the articles discusses the CSE curriculum. These types of articles "mainly present a new curriculum in their institution and elaborate on teachers and students' experiences." An example of this type of article is Garland (1994).
- Choose 5 if the article discusses program visualization.
- Choose 6 if the article discusses simulation.
- Choose 7 if the article discusses parallel computing, (e.g., Schaller & Kitchen, 1995).
- Choose 8 if none of the categories above apply.

DE8a. This variable is from Valentine's (2004) methodological review. (The quotes are all from Valentine.) Choose only one category, from the categories listed below.

1= Experimental:

If the author made any attempt at assessing the "treatment" with some scientific analysis, I counted it as an "Experimental" presentation. . . . Please note that this was a preemptive category, so if the presentation fit here and somewhere else (e.g. a quantified assessment of some new Tool), it was placed here. (p. 256)

Note if *experimental* was selected on DE8a, then DE9 should be *yes* and DE9a should be *no*. If DE9a (anecdotal) was *yes*, then DE9 should be something other than *experimental* — the assumption being that informal anecdotal accounts are not appropriate empirical analyses.

2. Marco Polo

The second category is what has been called by others "Marco Polo" presentations: "I went there and I saw this." SIGCSE veterans recognize this as a staple at the Symposium. Colleagues describe how their institution has tried a new curriculum, adopted a new language or put up a new course. The reasoning is defined, the component parts are explained, and then (and this is the giveaway for this category) a conclusion is drawn like "Overall, I believe the [topic] has been a big success." or "Students seemed to really enjoy the new [topic]". (p. 256)

3. Tools

Next there was a large collection of presentations that I classified "Tools". Among many other things, colleagues have developed software to animate algorithms, to help grade student programs, to teach recursion, and to provide introductory development platforms. (p. 257)

4. John Henry

The last, and (happily) the smallest category of presentations would be "John Henry" papers. Every now and then a colleague will describe a course that seems so outrageously difficult (in my opinion), that one suspects it is telling us more about the author than it is about the pedagogy of the class. To give a silly example, I suppose you could teach CS1 as a predicate logic course in IBM 360 assembler – but why would you want to do that? (p. 257)

5. Philosophy

A third classification would be "Philosophy" where the author has made an attempt to generate debate of an issue, on philosophical grounds, among the broader community. (p. 257)

6. Nifty

The most whimsical category would be called "Nifty", taken from the panels that are now a fixed feature of the TSP. Nifty assignments, projects, puzzles, games and paradigms are the bubbles in the champagne of SIGCSE. Most of us seem to appreciate innovative, interesting ways to teach students our abstract concepts. Sometimes the difference between Nifty and Tools was fuzzy, but generally a Tool would be used over the course of a semester, and a Nifty assignment was more limited in duration. (p. 257)

DE9. (human participants) Choose *yes* if the article reported direct research done on human participants – even if the reporting was anecdotal. Choose *no* if the authors did not report doing research on human participants. For example, if the author wrote, "the participants reported that they liked using the Jeliot program," then *yes* should be chosen. But, if the author wrote, "in other articles, people have reported that they enjoyed using the Jeliot program," choose *no* since the research was not done by directly by the author. (If *yes* go directly to DE9a. If *no* go to A9.)

DE9a. (anecdotal). Choose this if the article reported on investigations on human participants, but *only* provided anecdotal information. If *yes* on DE9 and DE9a, end. If *no*, on DE9a then go to A11 and mark A9 and A10 as -9. This might include studies that the author purported to be a 'qualitative study,' but mark *anecdotal* if there was not evidence that a qualitative methodology was used and the authors were just informally reporting their personal observations.

A9. (type of other) If the article did not report research on human participants, classify the type of article that it was. Choose 1 – *literature review* if the article was primarily a literature review, meta-analysis, methodological review, review of websites, review of programs, etc. Choose 2 – *program description* if the article primarily described a program/software/intervention and did not have even an anecdotal evaluation section. Choose 3 — *theory, methodology, or philosophy* if the paper was primarily a theoretical paper or discussed methodology or philosophical issues, policies, etc. For example, an article that discussed how constructivism was important for computer science education would go into this (3) category. Choose 4 – *technical* if the article was primarily a technical computer science paper. For example, an article would go into this category if it compared the speed of two algorithms. Finally, choose the (5) *other* category if the article did not fit into any of the categories above. Use category 5 as a last resort. (If categories 1,2 3, or 4, are chosen go to A11. Otherwise go to A10.) (Choose only one.) (This variable was derived from previous the methodological reviews: Randolph [in press], Randolph, Bednarik, & Myller [2005]; and Randolph, Bednarik, Silander, et al., [2005].)

A10. (other other) If you chose category 5 on variable A9, please write a description of the paper and describe what type of paper you think that it is.

REPORT STRUCTURE

In this section, which is based on the structure suggested for empirical papers by the APA publication manual (2001, *Parts of a Manuscript,* pp. 10-30), you will examine the structure of the report. Filling out the report structure is not necessary if it was an explanatory descriptive study, since this report structure does not necessarily apply to qualitative (explanatory descriptive) reports.

A11. (abstract) Choose 1 – *narrative* if the abstract was a short (150-250) narrative description of the article. Choose 2 – *structured* if the abstract was long (450 words) and was clearly broken up into sections. Some of the abstract section headings you might see are 'background,' 'purpose,' 'research questions,' 'participants,' 'design,' 'procedure,' etc. A structured abstract does not necessarily have to have these headings, but it does have to be broken up into sections. Choose 3 – *no abstract* if there was not an abstract for the paper.

A12. (introduce problem) choose 1 – *yes* if the paper had even a brief section that described the background/need/context/problem of the article. Choose 2 – *no* if there was not a section that put the article in context, described the background, or explained the importance of the subject. For example, you should choose *yes* if an article on gender differences in computing began with a discussion of the gender imbalance in computer science and engineering.

A13. (literature review) Choose 1 – *yes* if the author at least mentioned one piece of previous research on the same topic or a closely related topic. Choose 2 – *no* if the author did not discuss previous research on the same or a closely related topic.

A14. (purpose/rationale) Choose 1 – *yes* if the author explicitly mentioned why the research had been done or how the problem will be solved by the research. Choose 2 – *no* if the author did not give a rationale for carrying out the study.

A15. (research questions/hypotheses.) Choose 1— *yes* if the author *explicitly* stated the research questions or hypotheses of the paper. Choose 2 – *no* if the author did not explicitly state the research questions or hypotheses of the paper.

A16. (participants.) Choose 1 – *yes* if the author made any attempt at describing the demographic characteristics of the participants in the study. Choose 2 – *no* if the author did not describe any of the characteristics of the participants in the study. (Choose 2 if the author only described how many participants were in the study.) If *yes* go to A16a. If *no* go to A17 and mark *-9* in A16a and A16b. Please note that this refers to the participants that were used in the evaluation of the section, not about participants who participated in the program in general. If they did not describe the participants in the study, you do not have to go to a16a and a17a.

A16a. (grade level). Categorize articles based on the grade levels of the participants participating in the program. If ages, but grades were not given, use the age references below. (Grades take precedent over age when there is a conflict.) If 6, go to A16b; else go to A17 and mark *-9* in A16b.

- Choose 1 if the students were in pre-school (less than 6 years old).
- Choose 2 if the participants were in grades Kindergarten to 3 (Ages 6-9).
- Choose 3 if the participants were in grades 4 through 6 (ages10-12).
- Choose 4 if the participants were in grades 7-9 (ages 13-15).
- Choose 5 if the participants were in grades 10-12 (ages 16-18).
- Choose 6 if the participants were undergraduates (bachelor's level) (18-22 years old).
- Choose 7 if the participants were studying at the graduate level (master's students) (23-24 years old).
- Choose 8 if the students were post-graduate students (doctoral students) (25-30 years old).
- Choose 9 if the students were post-doctoral students (31 and over years old).
- Choose 10 if more than one category applies or if the category that is appropriate is not listed here.
- Choose 11 if it is impossible to determine the grade level of the participants.

A16b. (curriculum year). If 6 in A16b, choose the year (1-4) of the corresponding undergraduate computing curriculum that the article dealt with.

A17. (setting) Choose 1 – *yes* if the author made any attempt at describing the setting where the investigation occurred. *Setting* includes characteristics such as type of course, environment, type of institution, etc. Choose 2 – *no* if the author did not describe the setting of the study. This might include a description of participants who usually attended a course or a description of the organization that the author was affiliated with.

A18. (instruments) Choose 1 – *yes* if special instruments were used to conduct the study and they were described. (For example, if a piece of software was used to measure student responses, then choose 1 if the software was described.) Choose 2 – *no* if special instruments were used, but they were not described. Choose -9 – *n/a (not applicable)* if no special instruments were used in the study.

A19. (procedure). Choose 1 – *yes* if the author described the procedures in enough detail that the procedure could be replicated. (If an experiment was conducted, choose *yes* only if both the control and treatment procedures were described.) Choose 2 – *no* if the author did not describe the procedures in enough detail that the procedure could be replicated. For example, if the author only wrote, "we had students use our program and found that they were pleased with its usability," then the procedure was clearly not described in enough detail to be replicated and 2 (*no*) should be chosen.

A20. (results and discussion). Choose 1 – *yes* if there was a section/paragraph of the article that dealt solely with results. Choose 2 – *no* if there was not a section/paragraph just for reporting results. For example, choose 2 (*no*) if the results were dispersed throughout the procedure, discussion, and conclusion sections.

METHODOLOGY TYPE

In this section you will code for the type of methodology that was used. Since articles can report multiple methods, you can choose all that apply. (These methodology types were initially developed from Gall, Borg, and Gall (1996) and from the American Psychological Association's publication manual (2001, pp. 7-8). Explanatory descriptive and exploratory descriptive labels came from Yin (1988). The descriptions of variables listed below evolved into their current from Randolph (2005, in press); Randolph, Bednarik, and Myller (2005); and Randolph, Bednarik, Silander, et al. (2005).

M21. (experimental/quasi-experimental) If the researcher manipulated a variable and compared a factual and counterfactual condition, the case should be deemed as *experimental or quasi-experimental*. For example, if a researcher developed an intervention then measured achievement before and after the intervention was delivered, then an experimental or quasi-experimental methodology was used. Choose 1 – *yes* if the study used an experimental or quasi-experimental methodology. Choose 2 – *no* if the study did not use an experimental or quasi-experimental methodology. Note if the author did a one-group posttest-only or retrospective posttest on an intervention that the researcher implemented, choose experimental/quasi-experimental. The posttest in this case might be disguised by the term 'survey.'

AS5. (assignment) Use 1 when participants knowingly self-selected into treatment and control groups or when the participants decided the order of treatment and controls themselves. Use 2 when participants or treatment and control conditions were assigned randomly. (Also use 2 for an alternating treatment design.) Use 3 when the researcher purposively assigned participants to treatment and control conditions or the order of treatment and control conditions or in designs where participants served as their own controls. Also use 3 when assignment was done by convenience or in existing groups. This variable originally was based on Shadish, Cook, and Campbell's (2002) distinction between experimental and quasi-experimental designs. They have been pilot tested in Randolph (2005, in press); Randolph, Bednarik, and Myller (2005); and Randolph, Bednarik, Silander, et al. (2005).

M22. (explanatory descriptive) Studies that provided deductive answers to "how" questions by explaining the causal relationships involved in a phenomenon should be deemed as *explanatory descriptive*. Studies using qualitative methods often fall into this category. For example, if a researcher did in-depth interviews to determine the process that expert programmers go through when debugging a piece of software, this should be considered a study in which an explanatory descriptive methodology was used. Choose 1 – *yes* if the study used an explanatory descriptive methodology and choose 2 –*no* if it did not. This does not include content analysis, where the researcher simply quantifies qualitative data (e.g., the researcher classifies qualitative data into categories, then presents the distribution of units into categories.)

M23. (exploratory descriptive) Studies that answered "what" or "how much" questions but did not make any causal claims used an *exploratory descriptive* methodology. Pure survey research is perhaps the most typical example of the exploratory descriptive category, but certain kinds of case studies might qualify as exploratory descriptive research as well. Choose 1 – *yes* if the study used an exploratory descriptive methodology and choose 2 –*no* if it did not. Note: If the author gave a survey to the participants and the investigation did not examine the implementation of an intervention, then you should consider that to be exploratory descriptive survey research.

M24. (correlational) A study should be categorized as *correlational* if it analyzed how continuous levels of one variable systematically covaried with continuous levels of another variable. Studies that conducted correlational analyses, structural equation modeling studies, factor analyses, cluster analyses, and multiple regression analyses are examples of correlational methods. Choose 1 – *yes* if the study used an correlational methodology and choose 2 –*no* if it did not.

M25. (causal-comparative) If researchers compared two or more groups on an inherent variable, an article should be coded as *causal-comparative*. For example, if a researcher had compared computer science achievement between boys and girls, that case would have been classified as casual-comparative because gender is a variable that is inherent in the group and cannot be naturally manipulated by the researcher. Choose 1 – *yes* if the study used a correlational methodology and choose 2 – *no* if it did not.

M26. (IMPDET). Use this if not enough information was given to determine what type of methodology(ies) were used. If M26 was *yes*, then end.

Examples. A researcher used a group repeated measures design with one-between factor (gender) and two-within factors (measures, treatment condition). That investigation should be coded as an experiment because the researcher manipulated a variable and compared factual and counterfactual conditions (the treatment-condition within factor). The investigation should also be classified as a causal-comparative study because of the between factor in which two levels of a non-manipulatable variable were compared. Had the researcher not examined the gender variable, this investigation would have only been classified as an experiment/quasi-experiment.

A researcher did a regression analysis and regressed the number of hours using Jeliot (a computer education piece of software) on a test of computer science achievement. In addition, the researcher also examined a dummy variable where Jeliot was used with and without audio feedback. Because of the multiple regression, the investigation should be classified as correlational. Because of the manipulatable dummy variable, the investigation should also be classified as an experimental or quasi-experimental design.

A researcher gave only a posttest survey to a class after they used the intervention that a researcher had assigned. The researcher claimed that 60% of the class, after using the intervention, had exhibited mastery on the posttest. Since the researcher claimed that 60% of the class had exhibited mastery on the posttest because of the intervention, then the investigation should be classified as an experiment or quasi-experiment (in M21) that used a one-group posttest-only research design (RD2). (Had the researcher did a survey, but not measured the effects of an intervention, then it would have just been exploratory descriptive and not a one-group posttest-only experiment.)

[Go to M27 if M21, M23, M24, or M25 = 1. Else end.]

M27. (selection) Choose 1 (random) if the sampling units were randomly selected. Choose 2 (purposive) if the participants were purposively selected. (For example, if the researcher chose to examine only extreme cases, this would be purposive selection.) Choose 3 if the research chose a convenience sample or existing group. Choose 3 unless there is evidence for random or purposive sampling.

EXPERIMENTAL RESEARCH DESIGNS

If an experimental / quasi-experimental methodology was used, classify the methodology into research design types. Choose 1 for *yes* and 2 for *no*. If *no* go to 1i and mark the rest of the variables in this section as -9. These designs were originally based on the descriptions of designs in Shadish, Cook, and Campbell (2002) and in American Psychological Association (2001, pp. 23-24). They had been previously pilot tested in Randolph (2005, in press); Randolph, Bednarik, and Myller (2005); and Randolph, Bednarik, Silander, et al. (2005), except for the *multiple factor* category.

RD1. (designs) Choose 1 if M21 was marked as *yes*. If so, one of the following variables must be coded as a *yes*. If *no*, mark -9 in all of the following RD variables.

RD1a. (design?) Choose 1 if RD1 was marked *yes* but it could not be determined what research design was used. Choose *no* if the design could be determined and go on to RD2. If *yes*, go I1.

RD2. (post-only) Use this for the one-group posttest-only design. In the one-group posttest-only design, the researcher only gives a posttest to a single group and tries to make causal claims. (In this design the observed mean might be compared to an expected mean.) This includes retrospective posttests, in which participants estimate impact between counterfactual and factual conditions.

RD3. (post controls) Use this if the posttest with controls design was used. In the posttest with controls design the researcher only gives a posttest to both a control and treatment group. Put the regression-discontinuity design into this category too and regressions with a dummy treatment

197

variable into this design. (The independent T-test, regression with a dummy variable, or univariate ANOVA analyses might be used with this research design.)

RD4. (prepost only) Use this for the pretest/posttest without controls design. In pretest/posttest without controls design the researcher gives a pretest and posttest to only a treatment group. (Dependent T-tests might be used in this design.)

RD5. (prepost controls) Use this for the pretest/posttest with controls design. In the pretest/posttest with controls design the researcher gives a pretest and posttest to both a treatment and one or more control groups. (Independent T-tests of gain scores or ANCOVA might be used on these designs)

RD6. (repeated) Use this for repeated measures designs. In the group repeated measures design, the researchers use participants as their own controls and are measured over multiple points of time or levels of treatment. (Repeated measures analysis might be used in this design.)

RD7. (multiple) Use this for designs with multiple factors that examine interactions. If only main effects are examined, code the research design as a control group design (like the case in a one-way anova.)

RD8. (single) Use this for single-subject designs. In this design, a researcher uses the logic of the repeated measures design, but only examines a few cases. (Single-case interrupted time series designs apply to this category.)

RD9. (IMPDET) Use this if the author did not give enough information to determine what type of experimental research design was used.

RD10. (other) Use this category if the research design was well explained but were not RD2-RD8.

RDH. (posttest only highest) Choose 1 if the only research design was the one-group posttest-only design (i.e., if RD2 was marked *yes*, and RD3 through RD10 were marked *no*), otherwise mark *no*.

This construct behind this variable is whether a researcher compared a factual with a counterfactual occurence. It assumes here that the one-group posttest-only design does not compare a factual with a counterfactual condition.

[Go to Ii –measures.]

INTERVENTION (independent variable)
For this group of variables, choose 1 – *yes* if the listed intervention was used in the article and choose 2 – *no* if the intervention was not used. Choose all that apply. These intervention codes were based on codes that emerged in the previous methodological reviews: Randolph, (2005) and Randolph, Bednarik, and Myller (2005).

I1. (intervention) Choose 1 — *yes* if an intervention was used in this investigation. Choose 2 – *no* if an intervention was not used. There might be an intervention in an experimental/quasi-experimental study or in an explanatory descriptive study. But, there would not be an intervention in a causal-comparative study, since it is examines variables not manipulated by the researcher. Also, there would not be an intervention in an exploratory descriptive study (e.g., survey study) since exploratory descriptive research is described here as research on a variable that is not manipulated by the researcher.

[If I1 = 1, go to I2, else go to D1 and mark all I variables as -9.]

I2. (student instruction) Choose *yes* if students were given instruction in computer science by a human or by a computerized-tool. Otherwise, choose *no*.

I3. (teacher instruction) Choose *yes* if teachers were instructed on the pedagogy of computer science. Otherwise, choose *no*.

I4. (CS fair/contests) Choose *yes* if students participated in a computer science fair or programming contest. Otherwise, choose *no*.

I5. (mentoring) Choose *yes* if students were assigned to a computer science mentor. Otherwise, choose *no.*

I6. (speakers) Choose *yes* if students listened to speakers who are computer scientists. Otherwise, choose *no.*

I7. (CS field trips) Choose *yes* if students took a field trip to a computer-science-related site. Otherwise, choose *no.*

I8. (other) Choose *yes* if an intervention other than the one mentioned here was examined. Otherwise, choose *no.*

DEPENDENT VARIABLES

In this section you code the dependent variables outcomes that were examined. Choose 1 for *yes* and 2 for *no.* Choose all that apply. These dependent variables codes were based on codes that emerged in the previous methodological reviews: Randolph, 2005; Randolph, Bednarik, and Myller (2005).

D1. (attitudes) Choose *yes* if student attitudes (including satisfaction, self-reports of learning, motivation, confidence, etc.) were measured. Otherwise, choose *no.*

D2. (attendance) Choose *yes* if student attendance or enrollment in a program, including attrition, was measured. Otherwise, choose *no.*

D3. (core achievement) Choose *yes* if achievement in core courses, but not achievement in computer science was measured. Otherwise, choose *no.*

D4. (CS achievement) Choose *yes* if achievement in computer science was measured — this includes CS test scores, quizzes, assignments, and number of assignments completed. Otherwise, choose *no.*

D5. (teaching practices) Choose *yes* if teaching practices were measured. Otherwise, choose *no.*

D6. (intentions for future) Choose *yes* if what courses, fields of study, careers, etc, that students planned to take in the future were measured. Otherwise, choose *no*.

D7. (program implementation) Choose *yes* if how well a program / intervention was implemented as planned (i.e., treatment fidelity) was measured. Otherwise, choose *no*.

D8. (costs) Choose *yes* if how much a certain intervention/policy/program costed was measured. Otherwise, choose *no*.

D9. (socialization) Choose *yes* if how much students socialized with each other or with the teacher was measured. Otherwise, choose *no*.

D10. (computer use) Choose *yes* if how much or how students used computers was measured. Otherwise, choose *no*.

D11. (other) Use this category for dependent variables that are not included above. Otherwise, choose *no*.

D11a. (describe) Please describe the intervention if it was 'other.'

MEASURES

In this section you will code what kinds of measures were used to measure the dependent variables. For some measures you will note if psychometric information, operationalized as the author making any attempt at reporting information about the reliability or validity of a measure. Choose 1 for *yes* and 2 for *no*. These measures codes were based on codes that emerged in the previous methodological reviews: Randolph (2005) and Randolph, Bednarik, and Myller (2005). For subquestions, if the head question was *yes*, then the subquestion must be either *yes* or *no*. If the head question was *no*, then the subquestion must be *-9*. For example, if M3 was *yes*, M3a must either be *yes* or *no*. If M3 was *no*, then M3a must be *-9*.

M1. (grades) Choose *yes* if grades in a computer science class – or overall grades (like GPA) — were a measure. Otherwise, choose *no*.

M2. (diary) Choose *yes* if a learning diary was a measure. Otherwise, choose *no*.

M3. (questionnaire) Choose *yes* if a questionnaire or survey was a measure— this includes quantitative questionnaires that had open elements. However, if a survey had all open questions, call it an interview (m6). Otherwise, choose *no*.

> M3a. (ques. Psych.) Choose *yes* if psychometric information was given about the survey or questionnaire. Otherwise, choose *no*.

M4. (log files) Choose *yes* if computerized log files of students' behaviors when using computers was a measure. Otherwise, choose *no*.

M5. (test) Choose *yes* if teacher-made or researcher-made tests or quizzes were measures. Otherwise, choose *no*.

> M5a. (test psych) Choose *yes* if psychometric information was given about the test or quiz. Otherwise, choose *no*.

M6. (interviews) Choose *yes* if interviews with students or teachers was used as a measure — this also includes written interviews or reflection essays. Otherwise, choose *no*.

M7. (direct observation) Choose *yes* if researchers observed strictly operationalized behaviors. Otherwise, choose *no*.

> M7a. (direct psych) Choose *yes* if reliability information (e.g., interrater agreement) was given about the direct observation. Otherwise, choose *no*.

M8. (stand. test). Choose *yes* if a standardized test (in core subjects or computer science) was a measure. Otherwise, choose *no*.

M8a. (psych. stand) Choose *yes* if psychometric information was provided for each standardized test. Otherwise, choose *no*.

M9. (student work) Choose *yes* if exercises/assignments in computer science was a measure – this might include portfolio work. This does not include work on tests, grades, or standardized tests. Otherwise, choose *no*.

M10. (focus groups) Choose *yes* if focus groups, swot analysis, or the Delphi technique were used as measures. Otherwise, choose *no*.

M11. (existing records) Choose *yes* if records such as attendance data, school history, etc were used as measures. This does not include log files. Otherwise, choose *no*.

M12. (other) Choose *yes* if there were measures that were not included above. Otherwise, choose *no*.

M12a. (explain other) Explain what the other measure was, if there was one. Otherwise, choose *no*.

[go to F1.]

FACTORS (NON-MANIPULATABLE VARIABLES)
In this section you will examine the factors or nonmanipulatable variables that were examined. (If they were manipulatable – they should be mentioned as an intervention.) Choose 1 for *yes* and 2 for *no*. These factors codes were based on codes that emerged in the previous methodological reviews: Randolph, (2005) and Randolph, Bednarik, and Myller (2005).

F1. (factors) Choose *yes* if any nonmanipulatable factors examined. [If *yes*, go to F2; else S1 and F2-F8 are *-9*.] Otherwise, choose *no*.

F2. (gender) Choose *yes* if gender of the students or the teacher was used as a factor. Otherwise, choose *no*.

F3. (aptitudes) Choose *yes*, for example, if the researcher made a distinction between high and low achieving students. Otherwise, choose *no*.

F4. (race/ethnic origin) Choose *yes* if race/ethnic origin of participants was used as a factor. Otherwise, choose *no*.

F5. (nationality) Choose *yes* if nationality/geographic reason/ or country of origin was used as a factor. Otherwise, choose *no*.

F6. (disability) Choose *yes* if disability status of participants was used as a factor. Otherwise, choose *no*.

F7. (SES) Choose *yes* if the socio-economic status of students was used as a factor. Otherwise, choose *no*.

F8. (other) Use *yes* if a factor was examined that was not listed above. Otherwise, choose *no*.

F8a. (explain other). Explain what the factor was if F8 was marked as *yes*. Otherwise, choose *no*.

[Go to S1]

STATISTICAL PRACTICES
In this section you will code for the statistical practices used. Choose 1 for *yes* and 2 for *no*. You can check all that apply. These categories come from the *Informationally Adequate Atatistics* section of APA publication manual (2001, pp. 23-24))

S1. (quant results) Choose *yes* if quantitative results were reported. Otherwise, choose *no*.
[If *yes*, go to S2; Else end and all following S2-S7 are *-9*.]

S2. (inf. stats) Choose *yes* if inferential statistics was used. [If *yes*, go to S3, Else go S8 and S3-S7 are -9)] If *yes*, head questions must be *yes* or *no*. If the head question was *yes*, then the subquestion(s) must be *yes* or *no*. If the head question was *no*, then subquestions should be marked -9.

S3. (parametric) Choose *yes* if a parametric test of location was used. — "e.g., single-group, multiple-group, or multiple-factor tests of means" APA [2001], p. 23. [If *yes*, go to S3a, else go to S4]

S3a. (means) Choose *yes* if either cell means and (cell sizes) were reported or if means cell variances or mean square error and degrees of freedom were reported. Otherwise, choose *no*.

S4. (multi) Choose *yes* if multivariate types of analyses were used. Otherwise, choose *no*.

[If S4 if 1, go to S4a; else go to S5]

S4a. (means) Choose *yes* if cell means were reported. Otherwise, choose *no*.

S4b. (size) Choose *yes* if sample sizes were reported. Otherwise, choose *no*.

S4c. (variance) Choose *yes* if pooled within variance or a covariance matrix was reported. Otherwise, choose *no*.

S5. (correlational analyses). Choose *yes* if correlational analyses were done. — "e.g., multiple regression analyses, factor analysis, and structural equation modeling" APA (2001, p. 23.) Otherwise, choose *no*. [If *yes*, go to S5a; else go to S6]

S5a. (size) Choose *yes* if sample size was reported. Otherwise, choose *no*.

S5b. (matrix) Choose *yes* if a variance-covariance or correlation matrix was reported. Otherwise, choose *no*.

S6. (nonparametric) Choose *yes* if nonparametric analyses were used. Otherwise, choose *no*. [If *yes*, go to S6a; else go to S7]

S6a (raw data) Choose *yes* if raw data were summarized. Otherwise, choose *no*.

S7. (small samples) Choose *yes* if analyses for small samples were done. Otherwise, choose *no*. [If *yes*, go to S7a; else go to S8]

S7a. (entire data set) Choose *yes* if the entire data set was reported. Otherwise, choose *no*.

S8. (effect size) Choose *yes* if an effect size was reported Otherwise, choose *no*. [If *yes*, go to S8a, else end.]

S8a. (raw diff.) Choose *yes* if there wasa difference in means, proportions, medians reported. Otherwise, choose *no*. (Here authors just needed to present two or more means or proportions. They did not actually have to subtract one from the other. This is also includes what is called 'risk difference.')

S8aa. (variability) Choose *yes* if a mean was reported *and* if had a standard deviation reported? If a median was reported, choose *yes* if a range was also reported. Otherwise, choose *no,* unless a mean or median was not reported, then use *-9* here.

S8b. (SMD) Choose *yes* if a standardized mean difference effect size was reported. Otherwise, choose *no*.

S8c. (Corr.) Choose *yes* if a correlational effect size was reported. Otherwise, choose *no*.

S8d. (OR) Choose *yes* if odds ratios were reported. Otherwise, choose *no*.

S8e. (odds) Choose *yes* if odds were reported. Otherwise, choose *no*.

S8f. (RR) Choose *yes* if relative risk was reported.

S8h. (other) Choose *yes* if some other type of effect size not listed above was reported. Otherwise, choose *no*.

S8i. (explain) If S8 was marked as *yes,* please explain what the effect size was. Otherwise, choose *no*.

Coding Book References

American Psychological Association. (2001). Publication manual of the American Psychological Association, 5th ed. Washington, D.C.: American Psychological Association.

Cole, J. P. (1991). WHILE loops and the anology of the single-stroke engine. *SIGCSE Bulletin*, *23*(3), 20-22.

Dawson-Howe, K. M. (1995) Automatic submission and administration of programming assignments. *SIGCSE Bulletin*, *27*(4), 51-53.

Gall, M. D., Borg, W. R., & Gall, J. P. (1996). *Educational research: An introduction*, 6th ed. New York: Longman.

Garland, W. & Levsen, V. (1994). Information systems curricula in AACSB accredited business schools. *SIGCSE Bulletin*, *26*(2), 26-30.

Kinnunen, P. (n.d.) *Guidelines of Computer Science Education Research*. Retrieved November 29, 2005 from http://www.cs.hut.fi/Research/COMPSER/ROLEP/seminaari-k05/S_05-nettiin/Guidelines_of_CSE-teksti-paivi.pdf

Mattis, W. E., (1995). An advanced microprocessor course with a design component. *SIGCSE Bulletin*, *27*(4), 60-64.

Randolph, J. J. (2005). A methodological review of the program evaluations in K-12 computer science education. Manuscript submitted for publication.

Randolph, J.J. (in press). What's the difference, still: A follow-up review of the quantitative research methodology in distance learning. *Informatics in Education*.

Randolph, J. J., Bednarik, R. & Myller, N. (2005). A methodological review of the articles published in the proceedings of Koli Calling 2001-2004. In T. Salakoski, T. Mäntylä, & M. Laakso (Eds.), *Proceedings of the 5th Annual Finnish / Baltic Sea Conference on Computer Science Education* (pp. 103-109). Finland: Helsinki University of Technology Press. Retrieved March 19, 2006 from http://www.it.utu.fi/koli05/proceedings/final_composition.b5.060207.pdf

Randolph, J. J., Bednarik, R., Silander, P., Lopez-Gonzalez, J., Myller, N., & Sutinen, E. (2005). A critical review of research methodologies reported in the full-papers of ICALT 2004. In *Proceedings of the Fifth International Conference on Advanced Learning Technologies* (pp. 10-14). Los Alamitos, CA: IEEE Press. Available online:

http://ieeexplore.ieee.org/xpls/abs_all.jsp?isnumber=32317&arnumber=1508593&count=303&index=4

Schaller, N. C. & Kitchen, A. T. (1995). Experiences in teaching parallel computing – Five years later. *SIGCSE Bulletin, 27*(3), 15-20.

Shadish, W. R., Cook, T. D., & Campbell, D. T. (2002). Experimental and quasi-experimental designs for generalized causal inference. Boston: Houghton Mifflin.

Valentine, D. W. (2004). CS educational research: A meta-analysis of SIGCSE technical symposium proceedings. In *Proceedings of the 35th Technical Symposium on Computer Science Education* (pp. 255-259). New York: ACM Press.

Yin, R. K. (1988). *Case study research: Designs and methods,* (Rev. ed.). London: Sage.